# Philosophical Perspectives on Brain Data

Stephen Rainey

# Philosophical Perspectives on Brain Data

palgrave
macmillan

Stephen Rainey
Ethics and Philosophy of Technology
Delft University of Technology
Delft, The Netherlands

ISBN 978-3-031-27169-4        ISBN 978-3-031-27170-0    (eBook)
https://doi.org/10.1007/978-3-031-27170-0

This Palgrave Macmillan imprint is published by the registered company Springer Nature Switzerland AG.
The registered company address is: Gewerbestrasse 11, 6330 Cham, Switzerland

# Acknowledgements

I owe many thanks to Dr. YJ Erden whose close reading of the text and insightful commentary on both style and content helped me to boost clarity throughout. Thanks also to Prof. Neil Levy, whose notes helped me better conceive of the text from a reader's point of view. Their constructive engagement was much appreciated while completing the book, which is better for it. Thanks too to an anonymous reviewer at Palgrave whose comments helped polish the final draft.

# Contents

# List of Figures

## Brains, Data, and Ethics

# Introduction

If someone knew everything about your brain, could they know every-thing about you more generally? If they had all the facts about your brain, would they have all the facts about what you thought and believed? If all those facts were compiled in a supercomputer, could that machine *calcu-late* what you cared about and wanted for the world? To approach these kinds of questions, it is important to clarify some of the right sorts of philosophical questions that ought to attach to brain research and how it uses data about the brain. This involves examining the contexts in which brain data are emerging, in research, clinical, and consumer contexts and raising questions in each context.

This book brings together philosophical analysis and neuroscientific insights to develop an account of 'brain data,' in a specific sense of the term. It explores what brain data are, how they are used, and how we ought to take care of them. It clarifies complex intersections of philo-sophical and neuroscientific interest, presenting an account of brain data that is comprehensible.

While science in general, including neuroscience, may be said to work with 'data' in terms of observations, in this context brain data represent something specific. Emerging trends in neuroscience appear to make it possible to explain the mechanisms of the mind through sophisticated processing of signals recorded from the brain. Neuroscientists are particularly interested in 'decoding' the signals produced by brain activity, like neurons firing. Recording this activity from specific regions of the brain can provide details on how different brain regions are involved in processes to do with perception, memory, physical movement, and so on. Brain recordings are complicated, with many different overlapping signals captured in any given period of recording. The need for sorting these complicated ensembles of signal is one that brings challenges. Sorting the various superimposed signals captured in any given period of brain recording requires algorithms that can pull out desired signals from the rest of the noise. This might be done according to a range of interesting wavelengths, or signal power, or some other metric to locate a target set of signals among the muddled whole. Once processed, the signal produces *brain data*.

Brain data can be used in computer processes, like constituting massive databases, for creating models of brain functions. They can provide evidence for answering a research question in a field like cognitive psychology; they might be used to assess brain health or illness as part of a clinical assessment; they could be used to control software applications or hardware devices. It might also be used to make predictions about mental or behavioural dispositions. From brain data, explanations for cognitive and behavioural activity can be derived, with consequences for how we think of ourselves, how we conceptualise mental health and wellbeing, and how we think sensitive data about ourselves ought to be treated.

Once brain data are acquired, they can be stored and aggregated in a way that permits a sort of brain data science. Data science in general can be applied to *crunch the numbers* on different kinds of behaviour, like spending patterns in cities, passenger activity in public transport networks, student outcomes relating to study patterns. Applied to brain data, it is hoped that this kind of approach could explain how neural mechanisms correlate with cognitive and overt behaviour. This can be done using varieties of methods including Artificial Intelligence (AI), with algorithms classifying brain signals as relevant to this or that cognitive or behavioural activity. With the possibility of deep learning, the opportunity for brain data to fuel a 'discovery science' of the brain is also made possible. This approach to science doesn't start with predictions to be supported through evidence found in experiment, but rather seeks patterns suggestive of new knowledge latent in massive amounts of data processed often automatically. With algorithms going beyond classification and into detailed *predictions* about the dynamics of the brain, having processed swathes of brain data, such developments could serve to bolster learning about dynamic states of the brain. This opens the possibility of brain insights based purely in data analysis.

These developments will have ramifications for concepts of the brain, the self, and the mind. But while brain data may revolutionise various approaches to the brain, they will not provide answers to questions about how the brain and mind may be related; how brain processes and concepts of mental health or illness ought to interact; how evidence derived from brains ought to be considered alongside overt behavioural evidence, like testimony. A cache of brain data on its own won't answer questions about whether we ought to be happy with ways in which our selves might come to be viewed through the lens of a datafied brain. While not necessarily answering such questions, the growth of brain data and their processing will nonetheless raise them.

Where there are data, there are also questions of ownership, leaks, and worries about misuse. When we consider data on our brains, the stakes are high. They may be higher than associated questions regarding data derived from social media and other such technology-mediated sources. While those kinds of sources can provide powerful signs about our behaviour, interpersonal associations, preferences, intentions, and moods, brain data relate to what's inside the skull. Brains are a nexus of everything we can think, say, feel, and do. What data suggests about our brains might conceivably bear upon any dimension of our being. It might leave us—at least apparently—open to being read like a book.

*Philosophical Perspectives on Brain Data* investigates the sorts of claims that are raised in different contexts like research, clinical, and the consumer, and how they ought to be considered. In doing so, the issue of how much hype and how much reasonable hope we should have for brain data is brought into focus. This is an important issue to bear in mind especially because neurotechnologies could develop at a pace outstripping our capacity to understand what's new, what's at risk, and what may be to gain.

Questions in the area of brain data will affect clinical practices like psychiatry by modifying how clinicians and patients interact, and by introducing AI-based diagnostic and treatment strategies. It might serve to prompt reorganisation of clinical categories of mental health and well-being; it might come to play a role in hiring and firing, and political and legal contexts as important for determining the characteristics of workers, voters, or suspects and victims in trials. Despite this potential, brain data are not very clearly characterised, conceptualised, or anticipated in policy. The issues arising are vastly complicated, as yet little understood, but of high importance. This book opens a novel space for evaluating hitherto arcane areas of academic research in order to provide the necessary scope for understanding some real-world consequences of brain data in different contexts. These consequences will include personal, socio-political, and public health dimensions. It is therefore vital that they are understood before their impacts upon aspects of everyday life can be evaluated.

# Brains, Data, and Ethics

**Abstract** The brain is essential in everything human beings think and do, consciously or unconsciously, purposefully, or automatically. To understand how the brain works would seem to provide a way to understand how the processes it enables or supports work too. This ought to lead to greater understanding of how things like perception, cognition, and behaviour in general arise. Understanding how the brain works looks like a road to understanding how human beings work in a fundamental sense.

**Keywords** Neuron doctrine • Split-brain • Neuroethics • Neuro-data ethics • Non-reductionism

Pioneering figures in neuroscience have sought greater understanding of how human beings work especially through examining the material from which the brain is constructed. In the C19th Johannes Purkinje identified large, branching neuronal cells by investigating thin slices of brain tissue. Those cells now bear his name—Purkinje cells. Otto Friedrich Carl Dieters would exploit growing sophistication in microscopy to provide descriptions of nerve cells with unprecedented clarity, including their input and output strictures—axons, and dendrites. Joseph von Gerlach and Camillo Golgi each developed ways of chemically staining samples of nerve tissue to reveal their structural features with clarity. Santiago Ramón y Cajal drew striking images of neuronal cells, hand drawn pictures from close observation of cells stained with Golgi's method using silver nitrate. Golgi and Ramón y Cajal shared the Nobel Prize for medicine in 1906 for their efforts. This early work did at least three things for neuroscience. It demonstrated the importance of structural features in the brain, and that those structures could be observed and described. It also served to place the neuron at the centre of neuroscientific investigation.

The neuron's importance as the primary focus for investigating activity in the brain—the so-called neuron doctrine—has remained even through subsequent developments in the C20th and C21st. This has been at the centre of metaphors of the brain and the mind as a computer, for example. It has also fuelled research in artificial intelligence, wherein biological neuronal models have served as the basis for designing learning machines.

But besides structure, a greater understanding of the brain has emerged through concentrating upon its dynamics. Structure is one thing, but in living, active brains, those structures are at work in ways even more complex than the presence of structures could reveal. There are estimated to be between 80 and 100,000 neurons per cubic millimetre in the human brain. That makes for upwards of 85 billion neurons overall in the average brain. This incredible structural density is astounding in itself. But when considered in terms of function, the connections among neurons come into consideration. Each neuron has many synapses—the connections among neurons—meaning that for an average brain there are something like 100 trillion synapses. It is this vast network of brain activity, with electrical impulses triggered in neurons and sent via synapses to other neurons and other cells, that must be grasped if the active brain is to be understood.

While Ramón y Cajal's meticulous and visually striking work remains fascinating and compelling, it is clear that microscopy and sketching images is not a technique translatable into a context wherein the active brain can be captured. An over-emphasis upon structural elements of the brain might, for instance, lead researchers to believe that different cognitive and behavioural phenomena are the products of different types of structure themselves determined organically, i.e. that the form cells take constitutes their function. While many neuronal cells are involved in an array of processes in the brain, some with more regularity than others, it would be a mistake to think there are cells dedicated to specific tasks, with these tasks somehow laid out in a grand design. There is structural order, complexity, and plasticity in brain functioning. Looking for answers by observing structures will provide only partial truths, with only a certain kind of value where the active brain is concerned. Neurofunctionality cannot be captured by techniques that are predicated on the kinds of structural description possible in static pictures. Nevertheless, imaging remains a keystone of neuroscience, albeit through powerful technical means rather than cell staining, microscopic observation, and artistic study.

Contemporary techniques of imaging the brain can capture both structural and functional features of the active brain, and thereby add novel insights to an emerging neuroscientific account of how the brain works. They do this through the recording of different kinds of indicators of brain activity. In Positron Emission Tomography (PET), for one, the presence of specific radioactive isotopes injected into the bloodstream can be detected as blood moves through the brain. This can provide a map of the minuscule vessels that structure the cortex. Hemodynamics, for another, is the study of blood as it moves through the body. By examining oxygen use within blood in the brain, it can be deduced where metabolic activity is occurring. Active cells require oxygen, and so where oxygen levels drop in blood within the brain, those areas can be inferred to be active. This recordable activity is called the blood oxygen level dependent (BOLD) signal, which relates to differences in magnetic properties of oxygenated versus deoxygenated blood. Besides this, much of the brain's activity is electrical with neurons 'firing' meaning their release of an electrical impulse along nerve fibres. By recording the electrical activity of the brain, it is therefore possible to map neural activity. Hemodynamics relating to the BOLD signal can be imaged through functional magnetic resonance imaging (fMRI) scans of brains, while electrical activity can be mapped via electroencephalography (EEG). Using such techniques, experiments can be designed in which participants are be exposed to experimental stimuli, like visual or auditory cues. Through imaging the hemodynamic and electrical activity of the brain during specific experimental tasks undertaken by participants, areas of the brain can be correlated with kinds of cognitive and behavioural phenomena based on functional responses to those stimuli.

In accounting for the working of the brain, can we thereby account for the cognition and action? After all, what would it really mean to 'understand' the working of the brain without reference to these phenomena? Part of this task is to come to a view on how the brain relates to those most interesting dimensions of ourselves. The brain's centrality for everything in human perception, cognition, and action implies that understanding one and the other are intimately connected. Exploring how this is the case seems to provide tantalising clues to human nature in neuroscientific terms.

## Neuroethics: What It Is, and Why It Exists

Questions relating to how we perceive, how we know, and how we can act have *philosophico-anthropological* stakes in appearing to mark out humanity in some fundamental senses. Philosophical accounts of perception, cognition, and action have been pervasive since philosophy began. As something so central to human experience, questions about our senses, knowledge, and behaviour in particular serve to frame the kinds of creatures we are. They are dimensions of being that stand us in contrast with objects, or plants, or other animals. Questions of how the senses relate to knowledge, for instance, have provided philosophers with rich material for thousands of years. How our senses contribute to knowledge of the world, whether we come to know the world as it is, or as conditioned by our physical or rational capacities, has powered debate across and within empiricist, rationalist, realist, and anti-realist philosophical camps. Philosophers have for a long time taken on the mantle of reflective analysis in this area, where the nature of human beings and their experience is at stake. This is probably because one of the most notable aspects of humanity is the capacity for thought itself. Human beings think all the time and can think about everything they experience and do. Where they don't do this explicitly, perhaps because they are skilled in an area and need not re-think something every time, they can often reconstruct the thinking they would have thought and explain themselves after the fact. As human beings think so much, and the business of philosophers includes thinking about thinking, it would seem to make sense they would take on this role.

Much of thought involves deliberation among reasons for one thing over another. It is the bringing together of disparate resources in order to produce an answer to a question, or a decision to act, or an expression of taste. In any case, thinking is an important thread in the fabric of human experience. Without reasoning, human behaviour would be hard to explain as anything other than just the natural behaviour of object in the world. A dropped ball does not *consider* its acceleration toward the ground, *agree* or not with its being bound up by gravity, *lament* or *endorse* its subsequent rolling and coming to rest. The physical world operates on a basis of natural laws, causes and effects discernible through observation. In physical respects human beings too are simply bound up in natural causes and effects. But the role of thinking in human experience sets it apart as something worth further investigation.

Human beings in general can decide about what they want to do, what would be prudent and what would be a bad idea, what they like or dislike. The world of human experience thus looks like it has a rational basis as well as one neatly captured by natural laws: a basis not apparently discernible by observation, but better grasped through different forms of explanation. This seemingly dual nature has provided puzzlement for successions of philosophers, who have responded in ways including the idea that if humans are to be understood properly, we must be understood non-physically. Descartes is one culprit here. Other philosophers and cognitive scientists, including Paul and Patricia Churchland, and Daniel Dennett, have supposed that in time science will provide all the answers we will need to explain away the trickier parts of human experience in scientific language, certainly without appealing to immaterial notions. The Churchlands and Dennett represent a particularly strong physicalist view, it's worth noting, while a great many other philosophers lie somewhere in the middle ground of *non-reductive physicalism*. This view makes no appeals to the immaterial, but nor does it hold that the mind might be explained fully in the language of physical science. Explanation of the rational dimension of human experience is in large part centred on understanding *why* a person thinks or does what they do. To capture this, we may need concepts that are hard or impossible to reduce to scientific language. A little more will be said on this below. But at any rate, empirical sciences are not mute on the topics of human action, reasoning, or judgement. Patricia Churchland is especially clear about this (Churchland, 1989). In short, there aren't clear and obvious reasons why empirical sciences shouldn't have something to say about these areas of human behaviour and experience.

'Neurophilosophy' can be seen as a philosophical response to growing knowledge about how the brain works. As such knowledge grew, it became clear that philosophical accounts of perception, memory, cognition, and judgement could stand to learn from growing neuroscientific and psychological findings. For example, 'split brain' experiments carried out in the 1960s provided evidence that objective observation of human experience could perhaps claim better knowledge of that experience than the first-person perspective (Gazzaniga, 1998). These experiments undertaken by Roger Sperry, Michael Gazzaniga, and Joseph Bogen involved patients whose corpus callosum—the nerve tissue connecting the two cerebral hemispheres—had been severed, producing a largely 'split brain,' with only very little residual interconnectedness between the hemispheres. This sort of procedure represented a last resort means of halting epileptic seizures in those for whom other treatments failed. It did this by severely limiting the ability for electrical activity to progress from one hemisphere to the other, thereby preventing neuroelectrical overloads characteristic of seizure onset. Perhaps surprisingly, the procedure often had no noticeable behavioural or mental side effects, despite the widespread, literal dis-integration of intra-brain communication. Under carefully constructed experimental conditions, however, split brains could indeed produce some strange results.

From an experimental point of view, split brain cases offer the possibility of stimulating the hemispheres independently. There is some truth in the idea that brain function is 'contained' in different sides of the brain. But it's important not to think of this too coarsely. It isn't the case that the left brain and the right brain contain functions completely—the common idea of the *logical* left brain versus the *emotional* right brain is far too simplistic. But to the extent needed for significance in the split brain experimental case, the left-brain can be regarded as the *verbal* brain hemisphere, and the right as largely non-verbal. The left brain is *dominant* in this respect. The right brain is dominant with respect to spatio-temporal processing. In one split brain experimental setup, different images are placed in the visual field of each eye. The right eye, connected to the verbal, left hemisphere sees an image of a chicken foot. The left eye, connected to the non-verbal right hemisphere sees a picture of a snowy scene. Asked to point to images to test associations among different pictures, the experimental subject's right hand, controlled by the verbal brain, points to a picture of a chicken shed. The left hand, controlled by non-verbal brain, points to a shovel. The odd bit comes when the participant is asked to explain *why* they are pointing at the shovel and the chicken.

The left hemisphere, verbal brain is dominant in making the associations required to give explanations—explanations are verbal after all. We might think therefore the subject would be at a loss as to why they were pointing to a shovel. The verbal brain has no access to the image of the snowy scene, just the chicken foot. The chicken foot and the chicken shed are clearly associable. But the shovel is best explained by the snowy scene, not visible to the verbal brain. Yet when asked for explanations, the participant will say something like: "the shovel is for clearing the chicken shed." This is immediate, without hesitation, and accompanied by apparent sincerity. The observer can see what has happened clearly, and explain it in terms of the split brain, and the manipulated visual field being presented to the subject. They can also explain why the participant explained things as they did, in terms of the information available to them and the contrasting functions of the brain hemispheres.

The phenomenon has been described in terms of the left brain as an 'interpreter' which produces a story beyond the available information, evidence of a human drive to seek explanations in general (Gazzaniga, 2005). But the story presented by the subject is strikingly at odds with these other explanations. It seems the verbal brain very quickly assimilates the apparent strangeness of association and just as quickly explains it away to the best of its ability. What does this tell us about human perception, knowledge and experience? Does the experimental observer know, objectively, more about the participant's situation? By extension, does detailed knowledge of the brain allow generalisation of this finding: are human beings routinely making their own experience through swift and unconscious assimilation of information into satisfying narratives, as an interpreter module of the brain processes scant sensory data? Questions like these suggest that there are details about perceptions and knowledge that can admit empirical evidence, besides philosophical reflection and analysis.

It certainly appears from this split brain experiment as if neurofunctionality constrains the ability the subject has to assimilate their own experiences. There is also the clear possibility that an external observer may know more about these constraints than the subject herself. Philosophers have had longstanding interests in cases of confabulation—like the construction of false memories, or cases of sincere but erroneous testimony. The split brain case looks like a nice neuroscientifically-generated candidate for exploration in terms of confabulation too. Being able to demonstrate that central facets of human experience and behaviour can be challenged in the ways illustrated by this kind of experiment, philosophers might well be motivated to revisit classic problems in philosophy of mind, theories of knowledge, accounts of action and responsibility, on the basis of neuroscientific evidence.

The idea of 'oneself' as an atomic centre of consciousness is apparently challenged in the outline of the split brain experiment just described. It seems some kind of information-combining is going on among separable segments of the brain in accounting for the experiences undergone by the subject. This seems far from Descartes' idea of oneself in his *Meditations* as a thinking substance, examining thoughts as one might examine objects, and evaluating their clarity and distinctness. This revisionary impetus also applies to accounts of ethics. After all, judgement and rational explanation are central elements in accounting for good and bad behaviour in human beings, but the split brain experiment suggests judgement is constrained by neurofunctionality. This further suggests that we may not be as responsible as we think we are in various circumstances, or that we need novel accounts of responsibility that include neural evidence. Ethics needs to take into account how the brain constrains cognition and behaviour if it is to legitimately critique how people behave. Neuroethics recognises distinctive philosophical and ethical issues where neuroscientific experimentation in general is considered. This also includes questions engendered by an understanding of the brain (Levy, 2007). We can make this idea acute by centring upon the role of responsibility and imagining the split brain experiment having gone slightly differently.

Let us suppose that an experimental subject is shown a series of images under the constraints of the split brain setup just described. But rather than making a conclusion that a shovel is for clearing a chicken shed, they draw a morally reprehensible conclusion to their set of images. Let's imagine the right eye is shown images of some people dressed in attire regularly associated with a historically rival nation. Among the images they may point to is a map depicting that rival nation. The left eye is shown an image of a nail. Among the images they can point to is a hammer. The associations made between the images in this case are between the person, a map of their associated nation, and a hammer. The verbal brain must assimilate this information via the interpreter, not knowing about the left eye seeing the image of a nail. Now let's imagine their confabulated story explaining the association between the person in national attire, the map, and the hammer is as follows: "I'm pointing at the hammer because the people of that nation are violent." The xenophobic conclusion is uttered by the subject in relation to the experience they are undoubtedly undergoing. But are they *morally responsible* for this reprehensible statement? We know that this experimental setup is precisely designed to exploit puzzling dimensions of human perception and explanation. The experiential inputs are manipulated in this setup precisely to prompt associations among stimuli based on sparse information. We will know from an external point of view how the image is better explained in terms of what the non-verbal brain hemisphere is seeing via the left eye. But this person here and now has uttered something seemingly racist. On the face of it, they have made a judgement and reasoned this to be the most appropriate conclusion. But the experimenter has another explanation for the genesis of the utterance not available to the participant.

Overall, the situation seems to provide a puzzle as to what kind of knowledge was drawn upon by the participant—was it a relatively automatic, quick association, based on immediate visual imagery? Was it evidence of a deep-seated racist conviction? Something else? If we decide that this person has indeed been racist, then for consistency's sake we need to conclude that in the original example the participant really did understand the shovel as being for clearing the chicken shed, and that was why it was pointed at. But this appears to fly in the face of the evidence we have as an external observer, with our knowledge of the snowy scene. We might otherwise conclude that this second person isn't racist, but that part of the conditioning of an important part of them includes the racist themes evidenced in their interpretive activity. We might say they are subject to *implicit biases*. In this case we would have to conclude that the role of the interpreter was quite powerful in steering self-reports of experience. Self-explanation—and with it, self-understanding—in this case would appear to be substantially limited, with non- or barely conscious mechanisms creating the narratives that constitute experiences.

Were the second participant to withdraw their remarks, to refuse to endorse them, and to repudiate racist ideas after the fact, how would we then consider what was happening? In the first example, we considered the interpreter as a fast mechanism, aiming to make for coherent experience by going beyond incomplete information to present to the person a clearer narrative. In the second case, this 'clearer narrative' was offensive. Where that offensive narrative is disavowed, it might appear that this was *better judgement* re-asserting itself, having been momentarily overshadowed by a mere confabulatory mechanism in the interpreter. But it isn't clear that the repudiation by the second participant ought to be considered a more reasoned expression, as opposed to simply further work of the interpreter. It can just as easily be seen as making this new scenario, one wherein offence has been caused, more coherent from the first-person perspective.

We are essentially asking here what role judgement is playing in self-understanding and explanation. The less of a deliberate role it appears to play, the more that human experience can be said to be explicable according to empirical investigation potentially minimising self-reports. Judgement might be explained away, eventually, on this conception. But the more authoritative of a role it has, such as where we accept the post-experimental regrets of the apparent racist as sincere and more representative of that person, we are left with puzzles. Not least, we must ask when and how, and to what extent, judgement really is central to human experience. This has direct implications for thought's role in human experience, and far-reaching consequences for how human beings ought to consider themselves as more like objects, animals, or as self-determining and rational creatures.

The brain can be seen to constrain thought and action in philosophically interesting ways. We have no obvious reasons to exclude empirical data from any philosophical accounting for the brain. We can re-assess classically philosophical conundra about thought and action in terms of what we know of the brain. There are consequences to be considered from this, especially in terms of ethics where thought and action collide in highly value-laden contexts. At the heart of this is judgement, as the point where we apply reasoning to what we know and what we want to do. Exploring these issues in terms of brain data is now especially important. Increasingly, the brain is known via data. While 'data' have always been part of the methods of science—a datum is any observation, data their plural—there is a specific sense of 'data' here that I think leads to questions in the context of the brain.

Complicated methods of brain recording are used to understand the structures and functional properties of brains. In cases of brain signal recording, signals are correlated with active areas of the brain. Electrical activity, for instance, might be associated with specific neurons in a specific area of the brain. Neuroscientists might then make hypotheses and experimental paradigms to explore and examine those associations, in order to understand brain activity, and how it relates to, or covaries with, cognitive or behavioural activity. When signal recordings are processed and aggregated at scale, they can come to represent the brain in ways other than this. Aggregated brain signal recordings, processed so as to populate large sets, are the kinds of data I am considering here and throughout. These aren't only the 'data' of observation correlated with physical brains and their activity, but a step beyond, relating brain data with other brain data.

Neuroscience in this datafied vein moves away from what's in the skull and toward what's modelled by data science on a hard drive. Patterns in data come to serve as proxies for observed correlates between brains and behaviour. Through data intensive brain imaging techniques, there can be seen a move away of—say—looking at neuronal activity in a fine-grained sense. Through data-intensive techniques, understanding the activity of the brain becomes less focussed on *which neuron fired when*, and more on broad patterns of brain-wide activations representable only in silico. 'Which neuron fires when' is essential in techniques like EEG, but the information EEG studies have provided across multiple years and domains can be absorbed into a brain data context such that the details matter less than emergent predictive models.

The shift in emphasis towards patterns in data is a way to boost understanding through including computational approaches to the brain. How brain data do and don't represent actual brain signals, and how which signals are produced (selected from the overall hum of brain activity as relevant to the enquiry), is vital in accounting for brain data. But while such data may be far from the original signals in their original recording contexts, nevertheless predictive power continues to increase and fruitful instrumental intervention on the brain continues.

How brain signals are recorded, how they are processed, what data results all raise questions about data acquisition, transparency of processing, curation, ownership of the data, and fair or reasonable uses of it. Where data science coalesces with neuroscience, there is the possibility for a step-change in understanding the brain. But with this comes associated risks familiar to other contexts where data has taken centre-stage. Data can be inscrutable and can transgress received wisdom about the very manner in which scientific investigation is carried out. Just as knowledge of the brain prompted a novel discipline of neuroethics, the datafication of brain knowledge ought to prompt brain data ethics as a significantly novel means of understanding the constraints upon human cognition and behaviour. We now need a philosophical account of brain data. This is because data is central to understanding the brain, understanding the brain contributes to understanding human cognition and behaviour, and so understanding brain data contributes to understanding cognition and behaviour.

# Neuro-Data Ethics: What It Is and Why It Ought to Exist

Increasing understanding of the brain has prompted groups of philosophers to re-examine philosophical accounts of cognition and behaviour, leading to novel approaches in neurophilosophy and neuroethics. But contemporary understanding of the brain itself is increasingly constrained by complicated issues in data. Recordings from the brain promise direct disclosures of brain activity correlated with cognition and behaviour. These disclosures are 'direct' in not requiring any discussion with the agent involved. The data speaks for itself, to some extent, in providing a story about cognition and behaviour not reliant on first-person testimony. This would bypass potential confabulations, like those presumably concocted by the interpreter function in the split brain experiment. But the sense in which a brain recording can 'speak for itself' is metaphorical.

Brain recordings must be processed to make them useable, and to reveal underlying patterns of interest. In this way, the recording of brain activity must be communicated through data processing. Examples of this communication include PET and fMRI images, themselves based in statistical techniques signifying radioisotope decay and blood oxygen levels revealed through changing magnetic properties of blood. They may also be communicated through the outputs from EEG recordings, processed electrical graphs signifying cortical neuron activity correlated with cognitive and behavioural activity. The means of presenting the datafied brain is of great importance, as the decision made about which means of presentation, and how that specific presentation is made, have implications for what can be shown. The production and uses of data have a bearing on how objective their contents can be, and thereby implications for how we ought to think about their meaning. In order to formulate questions about brain data it must first be established to what 'brain data' refers, and thereby what 'datafication' means. The nature of the techniques, what they record, how this data can be processed, affects how the data ought to be considered.

The apparent strangeness of the situation evoked in split brain experiments suggests an objective story is possible concerning the first-personal reports people may give about their activities. Through expanding neuroscientific understanding of structural and functional features of the brain itself, we might illuminate how such features constrain judgement. But given the anomalies present in those first-person reports (e.g. the shovel for clearing the chicken shed), a way is needed to bypass those reports. It seems that the testimony of the researcher can only go so far in overriding participant reports. Puzzles remain about how properly to account for what happens in those cases. In the split brain experiment, the researcher can notice discrepancies, and speculate as to what's happening in terms of the role of the brain as interpreter. But little can be said about *why* this might be, beyond a general account of the split brain being responsible. For a deeper understanding, the workings of the brain need to be somehow laid bare, in order to ground scientific investigation.

The artistic renderings of neurons made by Ramón y Cajal showed the structures present in the brain in a very striking way. These are pictures of the brain in a conventional sense of the word 'picture'. But they don't give much scope to propose or to test hypotheses about the brain. This changes where data enters the scene. Rather than an artistic, visual 'recording' of the brain, neural activity can instead be recorded in terms of its known electrochemical properties. From recordings including fMRI and EEG, neural correlates of thought and overt behaviour may be deduced or predicted. These could be (i) taken as underlying any self-reporting and (ii) expected to be separable from whatever cognitive processes that led to the puzzling testimony concerning the chicken shed and the shovel. The data might be held to reveal the fundamental basis on which such confabulation ran. It might also provide answers as to whether the apparent xenophobe was subject to an automatic process or one evidencing deliberate activity. Data would appear to promise objectivity, and thereby clarity, about such processes. It could even be held by some to challenge psychological evidence surrounding human motivations with a data-driven, neuroscientific account.

Data promises to be objective and give unvarnished access to the real, underlying activity of the brain. This has been the promise of data in all sorts of other domains, including in genomics research, human resources, and the global economy. Through representing complicated areas of enquiry in data, patterns can be found that are held to explain the complicated areas. The reduction of complexity to the simplicity of patterns in data is a way to boost understanding. But brain recordings are technically complicated in themselves and require algorithms to create usable brain data. The final images of fMRI scans, for instance, are quite far from the original neural signal they represent. They don't resemble them in the way Ramón y Cajal's drawings resembled actual neuronal structures. Brain data are processed by algorithms and sifted using statistical techniques in order to present findings of use or interest for some specific goal. This is one point where the brain data domain overtly converges with AI, in the use of deep learning networks. How brain data do and don't represent signals, and how which signals are produced (selected from the overall hum of brain activity as relevant to the enquiry), is vital in accounting for brain data.

Recognising the need for high levels of evidence in general for scientific claims, data plays a growing role. Data appears to offer access to information on scales not available to individual researchers, labs, or even whole institutions. The promise of data includes that it replaces the impossibility of a census-like approach to questions where there are potentially large populations at stake. Rather than surveying an entire group on some particular issues, data allows the mediation of a wide ranges of observations. Imagine if all of the experimental results from top labs all over the world were uploaded to a database somewhere in a cloud database. From this aggregation of leading research data, larger results might be obtained than those which were sought in the smaller experiments each lab carried out. Moreover, other labs ill-equipped to carry out large experiments might tap into this trove of data and produce findings of their own, without having to have done the primary research themselves. This is already the ambition of brain data services like those provided by the European Commission's Flagship *Human Brain Project* partner platform, 'ebrains.eu'.

A quickening of research, and egalitarian levelling of the research field mediated in clouds of data might be anticipated for the future. Many data-scientists advocate the view that through amassing ever more data, from the sheer volume hitherto obscure knowledge can be gained through patterns detectable in data. These patterns are held to demonstrate relationships among the large ranges of observations collated in the dataset. The relationships provide a basis to hypothesise about what the data represent, or to conclude questions about them. From these data, associations among the observations can be made and novel insights extrapolated from them—insights perhaps not even considered during the collection of the data in the first place. This new knowledge can then be put to use in advancing the field. For example, following the work of Ivo Dinov and his group in the University of Michigan, demographic information might be aggregated into a large dataset, combined with medical observations, genetic information, and other meta-data, all collected from different sources for different purposes:

"Such data can be mined, processed and interrogated to extract specific human traits, biological dynamics and social interaction information, which ultimately may lead to tremendous benefits (social, biomedical, financial, environmental, or political)" (Dinov, 2016, p. 6).

These data will have been amassed for different reasons, or for no specific reason. From their amalgamation however, insights not available but for the processing of the combined data may emerge and allow for predictions about population-level health issues. No specific, time-consuming, focussed experiment needs to be carried out, but benefits for predicting health and illness, or prioritising funding, or producing targeted healthcare policies can be found through crunching data from masses of sources, at different scales, collected for a variety of reasons.

Using data analysis tools and approaches, relationships among heterogeneous data yields patterns that provide insights. But the data are not themselves neutral. Some might already be concerned about the picture described above, wherein from data mining *specific human traits, biological dynamics, and social interaction information* can be derived. Dinov's optimism about the uses of this data as the basis for far-reaching benefits may not be shared by others who are concerned about the *creepiness* of a data mining system predicting their traits. But given these traits are to be extracted from the data, one can only be as optimistic as that data's objectivity permits. After all, if one were to have a skewed, impartial, or biased dataset, one couldn't expect an accurate picture of objective reality to emerge from the inferences drawn from those data. As is quite widely discussed already, biased datasets produce biased predictions that can serve to make 'feedback loops' that perpetuate the kinds of issues that fostered their own creation (O'Neill, 2016). If the data are not objective in a suitable sense, we ought to be careful about how they are used. At the very least, we ought to expect some detail on the ways in which we might expect them to be skewed, impartial or biased. That is, we should expect to know something about data generation.

In order to select data sources, record them, populate and maintain datasets, and operationalise them with sufficient quality, expert knowledge is required. This means expert choices are made about what counts as relevant data for the object of enquiry. This also means choices about whether the data is usable in a technical sense, such as whether it is in a format usable by a particular piece of software, or storable on a particular piece of hardware. These are epistemic and practical constraints that are made prior to the population of a database. Similar choices will be made at other stages in the data analysis life cycle. In the case of neuroscientific data for example, the expertise of data scientists, software and hardware engineers, and developers will be needed to ensure quality technical systems and data structures. But neuroscientific expertise will also be required to identify relevant data and connections between data. Indeed, how and in what respects data are accurate to the neuroscientific phenomena they represent will be a question in need of close scrutiny. This should include philosophical and ethical scrutiny. As the data are curated, their usefulness carefully honed, with professional and technical choices included, the database ought to be considered carefully from a variety of angles. This is all the more important where it might be expected to yield predictions about *specific human traits, biological dynamics, and social interaction information.*

In a context of growing technological complexity in recording, classifying, decoding, and processing brain signals, brain data are becoming increasingly central as a basis for far reaching predictions, and correlations among brains, behaviour, and mental activity. This centrality is due not least to the ineliminability of computational approaches in accounting for the variety of signals available for capture from the active brain. Much as neuroscientific and technical curation of datasets is required for amassing brain data sets, so too is computational power. Brain activity operates at different timescales, and across different functional levels, simultaneously. The complexity of activity makes the acquisition of useful brain signals from the mass of information a forbidding task. AI can play a role in this part of the story. AI is capable of discerning order from within the complexity of neural activity, and thereby selecting relevant data. This is just like the processing and pattern finding from large datasets desired in data science generally, albeit carried out on the activity of brains rather than voluminous big datasets. Artificial neural networks (ANNs), themselves based on models of organic neurons, are especially apt at the kind of pattern recognition most useful here.

ANNs are essentially sets of algorithms that operate on datasets to discern patterns in those sets. All neural networks can do two main things: learn from examples, and generalise (Mitchell et al., 1993; Schölkopf, 2015). This makes them extremely useful for pattern-recognition in particular. From the recognition of patterns, applications can be developed that generalise from the form of many examples to novel instances of a scrutinised type. One simple example is a 'perceptron', which is a neural network used to classify input data into two sets (Butterfield et al., 2016). If we had a perceptron that was trained to recognise the letter 'x', for instance, we could give it various inputs which it would classify as x or not x based upon its training (see Fig. 1).

Inputs sufficiently similar to a target, according to prior assigned probability thresholds, are classified as members of the target set. A real system, unlike the very simplified instance in Fig. 1, would have many exemplars, and the training data would be many more examples than just one 'x'. But the principle should be sufficiently illustrated by the sparse example. It should be clear how, for example, were a letter 'O' to be input to the illustrative perceptron, the 4 segments into which it would be cut would be very unlike those of an 'x' and so the system would classify the 'O' as 'not-x' (see Fig. 2).

Neural recordings, taken at scale and processed by AI, can be classified in fine detail as similar or not to one another, in a range of degrees. This can be based in features of brain signals, like their frequency or how they

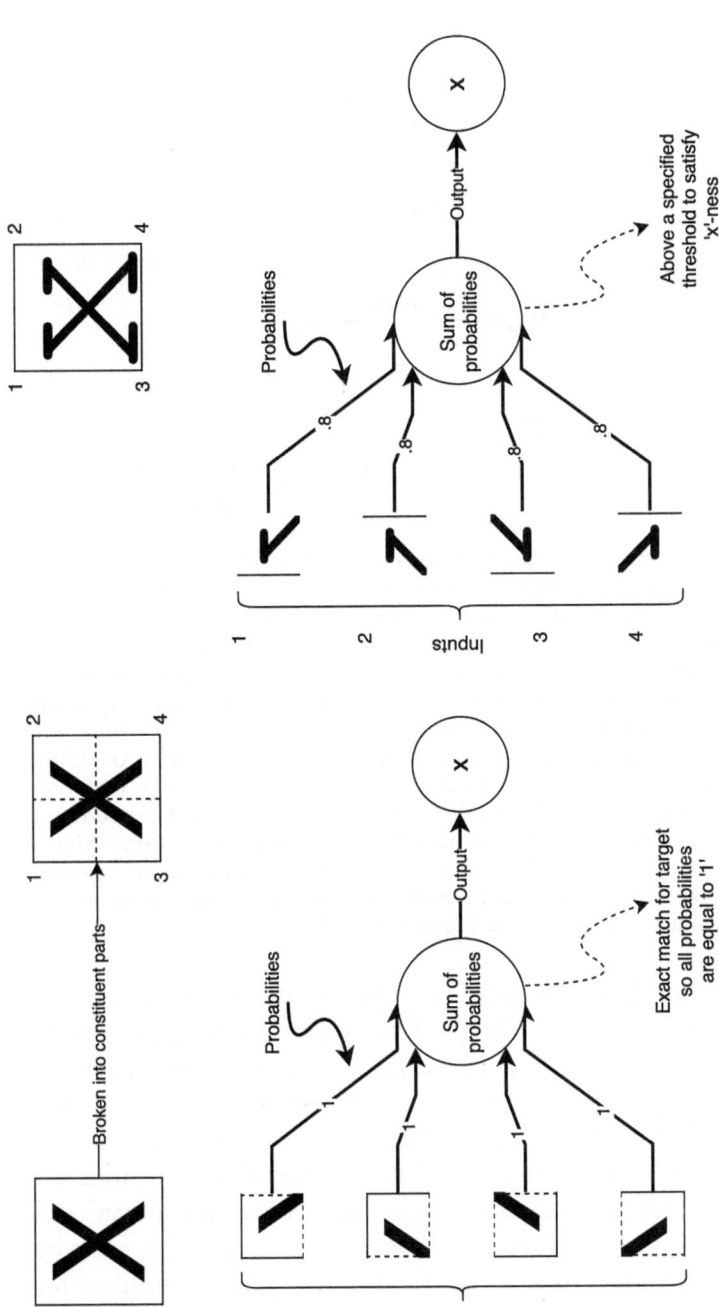

**Fig. 1** A very simplified outline for the principle on which a perceptron neural net operates. The target is an 'x' and the perceptron here will work on 4 nodes, so cuts the image into 4 parts. When presented with novel data, in this case represented by an 'x' in a different font, the new input is cut into 4 parts. Each of the 4 is then examined to determine its similarity to each of the 4 parts of the target 'x'. They are assigned a probability in terms of how well or not the input correlated with the target. In the first instance, the input is the target, so the probability is 1. In the second instance, each segment of the 'x' in a new font is supposed to be around 80% similar to the target 'x'. This is deemed sufficient for 'x'-ness according to a prior assigned probability threshold, and so the new 'x' is determined to be an 'x', and the perceptron has worked in classifying this novel input

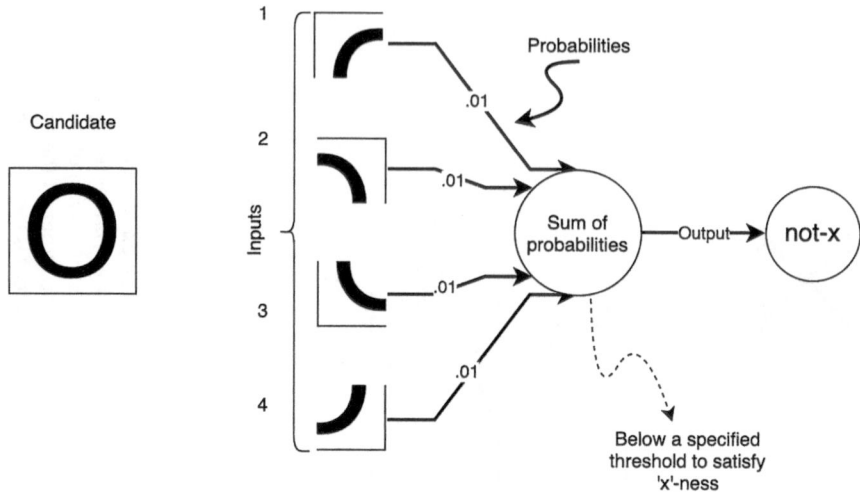

**Fig. 2** A candidate input is classified by the perceptron as 'not-x' because each of its segments falls below a threshold for determining x-ness. The probabilities are so low because each segment is very unlike those of the target (or any) 'x': they have curved contents, for instance, and so fill the input nodes differently to the straight lines of an 'x'. The kinds of principle illustrated in Figs. 1 and 2 seem far too basic to be of relevance to a complicated task like that of processing brain signals. But the complexity of a neural net grows extremely quickly, especially when 'layers' are added. Layers are extra steps between input and probability summing, which can feed back and forward, essentially communicating with one another. As layers are added to the single layer perceptron model outlined here, they are said to become 'deep' neural nets. The depth indicates layers. What's more, classification can be made far more complex as layers are added, making the work of the deep neural net more like pattern-recognition or discovery, not simply binary classification (impressive though that is)

co-occur with other signals, or how they repeat in longer timescales. This kind of sorting and pattern matching offers the promise of deeper understandings, or even explanations, of perceptual, cognitive and behaviour-related brain activity. Classifying brain activity allows order to be shown through data. This appears to confirm the role of datafication as revealing implicit order. But it should be noted that the objectivity here promised in data is not as clear as it might be supposed. Between the 'target' and the 'candidate' cases in the illustrated figures above, a standard must be set toward which AI processing tends. This is called 'ground truthing' for an AI system. In setting a ground truth, an artificial neural net is trained to recognise its target through exposure to many examples. A system can be trained by fixing the outcome, and then setting probability thresholds so that good candidates are selected that match the target. In the case of a visual representation of a letter and a perceptron, this is easily illustrated. But with the vastly more complex case of neural signals and deep learning networks, the issues surrounding ground truthing become equally complicated.

This is well illustrated by looking at a parallel case, centring on one of the largest medical AI datasets in use, known as "ChestX-ray14". As Kulkarni et al. (2020) discuss this dataset was used in a study to train an AI detection model called "CheXNet." This AI model would be used to carry out the functions of a radiologist, examining x-rays to diagnose problems in patients. AI is well-known as excellent in pattern recognition, so the tell-tale signs of problems of hernias or lung shadows in x-ray images would seem ideally suited for AI processing. In order to know what CheXNet should look for, the 'ground truth' had to be provided. Like setting the target 'x' in the example above, this is so that candidate inputs can be sorted according to required standards, and positive cases might be defined. Ground truthing here was based on a text-mining approach using radiology reports from experts. In the end, however, the model reproduced perhaps too closely the performance of human radiologists:

> "Intriguingly, CheXNet's performance mirrored human weaknesses in many respects; the algorithm had much greater accuracy in detecting hiatal hernias, a radiographically distinctive diagnosis, compared to pulmonary infiltration, which is frequently ill-defined" (Kulkarni et al., 2020, p. 63).

The notes from which the ChestX-ray14 dataset is comprised is itself rife with uncertainties. This is because diligent radiologists note their hypotheses, their diagnoses, their doubts, and so on. Diagnoses aren't always binary cases of examining an image and saying of it 'yes' or 'no' to the presence of an object of enquiry. While there is this necessary ground truthing step in training AI systems like CheXNet, there isn't one for real life radiology. There is training on cases, learning from colleagues, and gaining expertise through experience. The point worth noting here is that the data processing approach, and the datafication of radiological evidence, doesn't clearly demonstrate the drawing back of a veil to show objective reality. What is shown is the contrast between the modes of learning human clinicians participate in and those AI undergo.

Human learning in a very abstract sense can be seen in terms of predictions about outcomes, updated in light of results. But when we learn complicated things like how to pick a good wine from a wine list, or how to evaluate a newspaper's reporting, or whether it's a good time to interject in a conversation, these centre on reasons and judgement calls. The connections among our decisions in these, and myriad other such human contexts, are reasons. In principle, we could be asked to explain these connections and provide the reasons for why we acted as we did. This isn't necessarily the case for all of human behaviour, much of which appears to be relatively automatic, but in a great many important areas, we act on a hypothetically explicable, rational basis.

Artificially intelligent systems don't do this. AI connects input elements with probability calculations. The specific requirement of the learning AI system leads directly to a reproduction of human-level traits in diagnosis. But it has no reasons for any of its activity. This isn't just because the system is impersonal, but because its inputs and outputs are not rationally connected. They are solely about the likelihood, based in prior probability and ground-truthing, of a new input resembling an old one in some respect. This doesn't make the system unimpressive. But it does show how hopes of objectivity need detailed scrutiny, and evaluation. What sense, for instance, can we make of 'objective reasoning and judgement' compared with 'objectively calculable likelihood'? What's important for now is to bear in mind that we can learn lessons for brain data from other domains where data has already taken a central role in explanations and methods.

Besides medicine, across genomics research, practices in human resources, and the workings of the global economy, questions have arisen concerning the uses and limits of data. This has prompted wide-ranging discussions over its nature, ownership, fair use, and the ramifications of 'datafying' established practices (Hogarth et al., 2008; Mayer-Schönberger & Cukier, 2013; O'Neill, 2016). Brain data are a relatively new thread in this fabric. Investigating the ways in which brain data can be used inevitably raises questions not just about techniques of brain recording, but also how recording technologies converge with brain data processing. These include epistemic and practical questions, and considerations of relevance. They also bear upon the uses of technologies, and the extent to which processing is automatic, artificially intelligent, or opaque to analysis. Given signal sorting includes algorithms, and algorithms can generally be improved by 'training' them on greater amounts of data, there is a reinforcing loop of data-to-algorithm-to-data produced. Brain data can improve algorithms that already exist, while bootstrapping potentially new algorithms out of the brain data boom enabled by prior generations of algorithms. Novel signal sorting approaches can be enabled through the growth of this brain data ecosphere too.

All of this bears especially upon the supposed objectivity of brain data. Understanding neurofunctionality through a lens of brain data places distinctive constraints upon how that neurofunctionality is construed, as well as offering new approaches to brain research. Partly owing to the complexities of data processing in this overall picture, distinctive philosophical and ethical problems are raised. One useful philosophical way to think about this is in terms of the reduction of *mental properties* to *brain state properties* and *what makes the difference* when we think about action.

## Non-reductionism

Intuitively, one might say it is a mental state of desire that leads one to reach for a cup of coffee when a morning boost is needed. On a reductionist account, the physical neural correlates of desire cause the reaching. Talk of mental causation, on this account, could be replaced by talk of neural causation. Gaining an understanding of the neural then might be thought of as giving understanding of the mental, or explanation of how the mental *works*. Brain data might be thought to provide good insights to cognitive and behavioural performance where the facts about the brain and its activity are held to map closely onto facts about those performances, like desiring or reaching. For those like the reductionists already mentioned above, facts about cognitive and behavioural performances can be reduced to facts about the brain and its activity. For them, brain data could provide a turbo-charged way to account for a wide variety of mental phenomena. But a reduction like this isn't clearly possible.

One question that arises is the following: if a mental state is identifiable from one time to the next, but the neuronal state varies in identity, does it make more sense to describe the mental property or the neural property when talking about causality? Put differently: I can identify my desire for coffee from one instance to the next, and I can identify the group of neurons from which a subset will realise that desire physically in the brain, but I cannot predict which specific group of neurons might correlate with its realisation. Given it is the desire that makes the difference between reaching and not reaching for coffee (or reaching for something else instead) it would make sense to identify the neuronal correlates of the desire in terms of the desire—the mental state. It is this with respect to which the variety of neuronal realisers are similar, or classifiable together. If they can be batched together and called a 'brain state property', this can be best understood through their relation to the mental property of 'desire'. Bearing in mind some of the facts about brains, like their neuronal density and the manner in which groups of neurons correlate with mental states, it is not straightforward to *identify* brain states with mental states. At the very least, one and the same mental state (i.e. a mental state that can be identified from one instance to the next) will—as a matter of neuronal complexity—correlate with a variety of ensembles of realising neurons. There will be a many-to-one realising relation between neurons and mental states.

Mental properties are those properties that *make the difference* between one activity and another, since physical movement might be indistinguishable between reaching for coffee, reaching for tea, or flailing (they could be the same physical movements). What makes the difference is the desire state, as it is this which is satisfied by coffee reaching, thwarted by tea reaching, or frustrated by mere flailing. We could say of both brain state properties and mental properties that they each cause things to happen in the world: Brain state properties are causally efficacious with respect to neuroelectrical activity, nerve stimulation, muscle contraction and so on. Mental states are casually efficacious with respect to *reasons*, under a variety of descriptions. Brain state properties cause the kinds of bodily movements that realise reaching for coffee. But mental properties are those that identify the bodily movements as significant, for example, as between reaching for coffee, reaching for tea, or merely flailing an arm. Without reference to the desire for coffee specifically, the grabbing of tea or mere flailing can't be understood as leaving that desire unfulfilled.

Another way to think of the difference-making nature of mental properties over brain state properties comes in the idea of *prediction*. Let's say the processing of my brain data allows a neuroscientist to predict I am about to choose a beer, having been offered a choice between a beer and a soda water. That my hand is set to move beer-ward can likely be predicted, under controlled circumstances in a lab, from the properties identifiable in the brain connecting with onward motor-neuron activity relating to my limbs, etc. But do we conclude thereby that brain state properties have indicated a mental property—a *preference* for the beer? This is not signalled in the brain without further interpretation. Why I might reach for the beer could be through a preference for beer, a hatred of soda water, preferring things on the left, having a rule that I drink only beer on a Tuesday, or delight in random choices. These further descriptions of mental properties are what make the difference in characterising the act in question. The reduction to brain states appears to miss this.

Brain data compounds this issue for reductionism of the mental to the physical brain. Besides the reduction to the physical, brain data (as I am using the term) amalgamates, generalises, and abstracts away from any specific brain. This means that over and above the question of whether this ensemble of neurons correlates with this mental state or that, there is the wider issue of whether brain data can be said to refer to any particular neuron or brain at all. While brain data might furnish a detailed model of the brain and its dynamics, it will not relate to any specific brain and it will be far from anyone's reasons for thinking or acting. This raises questions about data collection, processing, ownership, and use, on top of existing questions about how observable brain activity relates to brains and to cognitive or behavioural action. In the face of such complication, the best way to find out about a person's actions and their mental states, might be to ask them.

There is a vast literature on non-reductive physicalism and related questions (cf, for example, Davidson, 2001; Pernu, 2018; Eronen, 2020), and the outline of a non-reductionist strategy here only begins to illuminate that position. Nevertheless, it is a very important perspective to bear in mind as the facts, ambitions, possibilities, and applications of brain data are examined further in what follows. It might be that, through brain data, we gain incredible advances in understanding the brain and its activity but acquire no particular insight on the mind or on what makes people tick. There is no clear way provided by brain data analysis to account for a distinction between reasoning as a cognitive activity, and brain-state transition as a neurofunctional occurrence based in probabilities. Constraints upon the potential uses of brain data are particularly pertinent in light of how these questions are answered. This is important in terms of developing a picture of distinctive philosophical and brain data-ethics problems where applications in research, health, or consumer markets are envisioned.

# Bibliography

Andrew Butterfield, Gerard Ekembe Ngondi, and Anne Kerr (eds) (2016) A dictionary of computer science. 7th edn. Oxford: Oxford University Press.

Churchland, P.S. (1989) Neurophilosophy toward a unified science of the mind brain. MIT Press.

Davidson, D. (2001) Essays on actions and events. 2nd edn Oxford; New York: Clarendon Press; Oxford University Press.

Dinov, I.D. (2016) 'Volume and value of big healthcare data', Journal of Medical Statistics and Informatics, 4(1), p. 3. doi:https://doi.org/10.724 3/2053-7662-4-3.

Eronen, M.I. (2020) 'Interventionism for the intentional stance: True believers and their brains', Topoi, 39(1), pp. 45–55. doi:https://doi.org/10.1007/s11245-017-9513-5.

Gazzaniga, M.S. (1998) 'The split brain revisited', Scientific American, 279(1), pp. 50–55.

Gazzaniga, M.S. (2005) 'Forty-five years of split-brain research and still going strong', Nature Reviews Neuroscience, 6(8), pp. 653–659. doi:https://doi.org/10.1038/nrn1723.

Hogarth, S., Javitt, G. and Melzer, D. (2008) 'The current landscape for direct-to-consumer genetic testing: Legal, ethical, and policy issues', Annual Review of Genomics and Human Genetics, 9(1), pp. 161–182. doi:https://doi.org/10.1146/annurev.genom.9.081307.164319.

Kulkarni, S. *et al.* (2020) 'Artificial intelligence in medicine: Where are we now?', Academic Radiology, 27(1), pp. 62–70. doi:https://doi.org/10.1016/j.acra.2019.10.001.

Levy, N. (2007) Neuroethics. Cambridge; New York: Cambridge University Press.

Mayer-Schönberger, V. and Cukier, K. (2013) Big data: A revolution that will transform how we live, work, and think. Houghton Mifflin Harcourt.

Mitchell, R.J., Bishop, J. and Low, W. (1993) Using a genetic algorithm to find the rules of a neural network. doi:https://doi.org/10.1007/978-3-7091-7533-0_96.

O'Neill, C. (2016). *Weapons of math destruction, how Big Data increases inequality and threatens democracy*. Penguin.

Pernu, T.K. (2018) 'Mental causation via neuroprosthetics? A critical analysis', Synthese, 195(12), pp. 5159–5174. doi:https://doi.org/10.1007/s11229-018-1713-z.

Schölkopf, B. (2015) 'Learning to see and act', Nature, 518(7540), pp. 486–487. doi:https://doi.org/10.1038/518486a.

# Research Contexts

**Abstract** Especially since the scientific revolution of the seventeenth century, a vibrant enthusiasm for the sciences has led to increasing faith in empirical methods of learning. At its best, this tendency pits speculation and curiosity against careful scientific investigations of physical reality. At its worst, it becomes a narrow fixation on specific and reductive ways of understanding reality. This latter sort of scientism is unfortunate, whereas science itself contains huge optimism about reality and our understanding of it. The sciences seem to give great explanatory power, being able even to explain the most fundamental dynamical laws of micro-causal interactions right up to macro-scale effects. To the early scientist and learned community at large, the sciences could reveal and explain the hidden designs of nature, or to usurp the supposed role for a divine design entirely, depending upon how one viewed the results of Newton, for example.

**Keywords** Scientific revolution • Datafication • Algorithms • Artificial intelligence • Dashboarding

What was notable about the scientific approach as it was seen by its earliest proponents was its unblinking, coldly analytical approach to *the facts*. Through experiment, hypotheses could be tested and shown to be well evidenced. They could be verified and falsified. New data could be used as a basis for further hypotheses. It must have seemed to many early scientists that no matter what the phenomenon under scrutiny, a full explanation could be gained of it using scientific methods. As explanation followed explanation, a general demystification of natural processes must have seemed possible. The world, revealed by means of science, seems to be explanatorily closed, i.e. every phenomenon in the world looked as if it was completely explicable in terms of other worldly phenomena. Furthermore, since the results of science were also repeatable and predictable regardless of the attitudes or opinions of experimenters, it looked as though the scientist had discovered a means by which to discover the true nature of nature itself, unhindered by bias or interest. So, in science it seemed objective truth was possible.

Fast-forward to the present, and the ambitions for objective knowledge borne by the sciences remains real, but more nuanced. Besides immutable facts the sciences—broadly construed—aim for discoveries relating to processes and probabilities. The physics of the past, for example, has yielded to recognising non-linearity and disjunction between micro and macro scales, as between quantum and relativistic physics. The reality revealed through science is no less real for being more complex than imagined, but it is sought through methods removed from disciplined observation, and instead planted more firmly in mathematical methods. The world as we know it through scientific investigation is as much revealed through calculation as observation. Data, in short, is what reveals scientific truths about the world at large. As with science in general, so too for the sciences of the brain. From the observational approaches of Ramón y Cajal, a step change is seen in contemporary neuroscience, with the brain revealed in unprecedented detail. This is achieved through a variety of recording techniques and revealed through processing swathes of data.

The brain as it is known via neuroscientific research includes the kinds of imaging available from PET scans, fMRI, EEG and others. In line with the neuron doctrine associated with Ramon y Cajal's analyses, neuron firings are still considered very central to understanding the brain and its activity, notwithstanding all the complexity afforded by having also to consider, for example, the chemical activity of neurotransmitters, and the physical constraints of neural structures. Neuron firings represent the brain *doing things* in response to external stimuli, or to other neuronal firings. These modes of recording brain activity throw light on the structures and functions of brain areas quite widely construed, right down to individual neuronal operation.

With structural and functional data gained from the brain, those data can be correlated with cognitive and behavioural activities undertaken by participants in research tasks. These might include things like observing specific images, calling words to mind, or responding to sounds. From the information collected during such tasks, relations among cognitive and behavioural activity and areas of the brain can be discovered. For instance, speech comprehension can be connected with characteristic neuroelectrical activity (Cheour et al., 2001). The recognition of speech in a native language is known to correspond with different brain activity than speech understood in a language learned later in life. A 'mismatch negative' (MMN) occurs when the brain manifests a broadly negative electrical polarisation upon the perception of a familiar sound. MMNs can be recorded by EEG. Knowledge of differentiation among neural response patterns to language perception can be used to explore types of language problems. It can be used to determine whether comprehension problems are cognitive or articulatory, for example, in cases of aphasia. If an MMN can be detected and thereby indicate that a sound is familiar, but verbal responses are missing, this might indicate that a motor problem is the cause of aphasia. By contrast, if the MMN is missing, this would suggest that the sounds being heard were not familiar to the listener, perhaps indicating a comprehension problem. These kinds are distinguished in the brain's electrical activity, when it occurs in response to speech, and where it occurs in the brain.

This is just one example of how insights to brain activity can open doors to detailed knowledge about hitherto obscure processes relating to human behaviour and understanding. This paradigmatic approach to determining brain states in relation to stimuli, is apparent throughout neuroscientific experimentation. An emerging thread in this research seeks not to find these kinds of states and their varieties of correlations, but instead looks for the dynamics of brain activity. In these whole-brain computational modelling approaches, 'brain states' are less like time-slices of neuronal firings, and more like models of tendencies among regions of brain activity and their coupled interactions. Data is central to all of the brain representations so far mentioned, but in this dynamic modelling variety data plays an immense role. The complexity of neural activity it seeks to capture can only be shown via computer models operating on vast arrays of data. In this way, data can be seen as vital to neuroscientific discovery, which means that conceptual and ethical questions arising around data will have implications for neuroscientific discovery. Once more, these can relate to epistemic, and practical choices made about brain recordings, and wider questions about why brain activity is considered essential for a question to be addressed.

The brain appears central to all of our cognitive and behavioural activity, and so revealing its operations would appear to be a kind of skeleton key for understanding humans in a very far-reaching sense. But there is an implicit possibility that the brain itself might be considered not just central, but the only target worth investigating. In the pursuit of neuroscientific knowledge, we might slip into thinking that knowledge of the brain equates with knowledge of the human. This reductionism would downplay wider factors relevant to human lives as they are lived—personal, social, political, historical dimensions of experience—in pursuit of facts about the brain and its working. Can we be sure that, in the excitement of a productive, datafied neuroscience, we will remember this is just one way of studying human cognition, behaviour, and so on?

Understood in terms of function, structure, neuronal activity, or in terms of patterns latent in troves of data derived from brains, differences will emerge regarding how we think of the brain. Where data takes centre-stage, data-reductionism emerges as an important point requiring scrutiny. As a device to explore the ramifications of brain reductionism, this chapter will imagine a scenario in which the brain comes to be considered a dashboard for the mind, and thereby for the cognition and behaviour of the person. This device, in which the brain is imagined as a kind of *control panel* for cognition and behaviour, will serve to illustrate some of the conceptual and ethical considerations that ought to attend neuroscientific investigation.

# Data in Neuroscience

Brain data can serve to produce images of brain function and structure and can provide a basis to predict cognitive and behavioural characteristics thereby. Brain data are derived from recordings of brain activity of varying sorts. Brain activity can be inferred from factors including from structural makeup via PET scan, BOLD signals, using fMRI, and from electrical activity as captured by EEG. Each mode of signal capture has benefits and limits. While it has excellent spatial resolution, for example, fMRI works on quite long timescales when it comes to brain activity. This means that while it shows blood oxygen use in clearly defined areas, it does not approach real time brain activity and so fast brain processes are not visible by this method. And that's not to mention the sheer complexity of an MRI machine, the cost, and the expertise involved in simply setting one up and using it. PET scans, meanwhile, include all this technical complexity and amplify it by involving specific kinds of radioisotope that decay quickly, meaning PET scanning requires a nearby nuclear chemistry lab. While they are of great use in showing structural features of the brain, PET scans can have poor spatial resolution due to inherent characteristics of the radioisotopes used, but also owing to their sensitivity to movements made by the subject.

This is all vastly complex, and computer processing of the signals involved in order to construct usable data is essential. Moreover, once captured these data must be processed statistically in order to determine the relative values of brain signals to one another, and against reference datasets. This is required in order to answer a research question, or otherwise put the data to work. This is then the basis for the sorts of images we are used to seeing, that neatly depict areas of the brain that are 'lit up' in relation to some described stimulus or activity. That these pictures are not images in the same sense as photographs but the result of complex signal capture, processing, and statistical calculation is easy to forget. But this is important, as it demonstrates such images are based in data, only visualised for ease of presentation. The raw data of a photograph is light bouncing off an object. The raw data of an fMRI image is an array of data.

Compared with fMRI, PET, and some other brain recording techniques, it is relatively easy to detect and record the electrical activity of neurons using EEG. This technique is highly portable and requires only fairly simple equipment. Commonly, EEG electrodes sit on the surface of the scalp, embedded in headgear, and record electrical activity occurring in the brain by detecting electrical fields the signals generate. This electrical activity is ubiquitous, and so a signal will easily be recorded from anywhere on the scalp using even the most basic equipment. But what to do with such a recorded signal is again a complicated affair.

Whilst EEG provides excellent temporal resolution in being able to operate on a scale of milliseconds, it has poor spatial resolution. Signals recorded come from throughout the brain and are transmitted through brain tissue, the skull, and the skin. From scalp surface recordings, processing must discern among a variety of signals jumbled with one another. For instance, one EEG electrode may detect signal from many neurons, at different locations and depths within the brain. Meanwhile, two different EEG electrodes may detect signal from the same neuron whose activity occurs within each of their range. Raw signal must be filtered, amplified, classified, and related to specific stimuli or other signals. These complexities must be factored in to the processing of brain signals to produce brain data that can be used for some purpose. This can be done with varying levels of success by applying mathematical techniques to signal recordings in order to separate signal sources and types from one another. This can involve processing signals in terms of time frequencies to aid in locating the brain region from which they originated, or according to analysis of signals to reveal specific characteristics known to have interesting properties. But in any case, as the specificity of the use of signals increases and purposes becomes narrower, the complexity of processing grows. This again places data centrally in the story. So, while EEG represents an easy mode of recording brain signals, it presents at the same time another locus for datafication as computerised processing becomes essential.

Data, and data processing, are at the heart of each of these brain signal recording approaches. Examining neuroimaging based in brain data has in-built limits based in the nature of the data curation, processing, and related factors. Unlike the limits in a photograph, such as not seeing detail in blurred objects or around corners, these limiting factors are not a natural limitation but are constructed. Despite the clarity of presentation possible by way of fMRI and PET images, or neat neuroelectrical graphs, they may yet serve to hinder understanding of the brain and its relations with cognition and behaviour. This is largely owing to a 'localisation problem' in neuroscience. This is essentially the fact that cognitive and behavioural phenomena, while too easily correlated with specific parts of the brain, include complex, brain-wide activity, and wider embodied and acculturated human experience (Rose, S., in Choudhury & Slaby, 2016, Chap. 2). Neuroimaging can encourage us to think of experiences like anxiety, or seeing the colour blue, thinking about objects, or being in one mood or another as things in the brain. But this attempt to localise the complex items of thought and experience in the brain doesn't stand up to scrutiny. Brain data plays a central role in this difficult area.

Owing especially to the datafication process, it might be easy to reduce the brain to a set of locations that function mainly as behavioural precursors, or as origin points of cognition and action. Datafication can encourage this reduction by offering an ordered perspective on the complexity of the brain. It might appear to underlie the open question of how to relate folk psychological talk of cognition and behaviour, and neuroscientific accounts. The data might appear to speak for themselves, and so to provide ultimate truths about these phenomena. The sophistication of data science bolsters this perspective through the deployment of computer technology capable of discerning patterns among data it would be practically impossible for a human scientist to see. The possibilities for processing brain signals would be limited where little computational power was available. Disentangling jumbled brain signals is complicated and requires an array of algorithms to be involved (Bashashati et al., 2007; Lotte et al., 2018). Algorithms are computational routines that can crunch huge amounts of data toward specific goals. This is especially useful for pattern detection and, in terms of brain recordings, for recognising patterns of activity that are characteristic of specific brain regions or functions (e.g. the visual system, or intending to speak).

The emphasis upon data and its processing offers a novel guise for an established mode of scientific endeavour in empiricism. Data appears as an objective dimension of brain activity, and thereby of cognition and behaviour in general. These data can be revealed by scientific methods centred on analysis of the active brain. Given the role of algorithms, moreover, in being able to detect patterns indiscernible to the human scientist, datafication seems to turbo charge empiricism. The role for theory is apparently diminished through using data science techniques— patterns that appear in troves of data are objectively discerned by algorithmic processing whether they are being sought out or not. One foundation for this approach comes from 'compressive big data analytics,' (CBDA) whereby huge datasets can be automatically sampled and sifted simply to find patterns and associations, without the need for prior formation of questions to be answered (Marino et al., 2018). In other words, by letting algorithms loose on troves of data we can discover patterns we didn't know it was worth looking for, and from those patterns derive objective knowledge *we didn't know we didn't know.*

Some have termed this the 'end of theory' and it has found advocates who would see data as ushering in a new era of discovery-led scientific research (Anderson, 2008; Toga et al., 2015). In terms of the brain, this would mean all of the recordings that are made of signals, across any number of experimental contexts, might be put into one big set and processed automatically. It would seem as if one could just put all the data together, press a button, and have data analytics highlight what's worth investigating further. If the patterns are in the data, after all, it would seem as if they are somehow in the structure of the cognitive and behavioural activities that created the signal recordings. Data processing looks like a way to unlock the secrets of the brain, and thereby make the human mind newly legible.

This is not a universal view, however, even beyond those who would resist reductionism in general. Counterpoints come especially from those working on the area of Big Data. For these critics of the data-centric utopian view theoretical choices and consequences are everywhere in the course of producing the tools and methods that collect data in general. Ditto when it comes to the interpretation of data. It is increasingly difficult to draw a clear line between data-driven and theory-driven science, as novel approaches and expectations emerge (Kitchin, 2014). Rather than data liberating some very general conception of empiricism in a new age of objectivity, instead the turn to data puts novel pressures on researchers to scrutinise their own role, as well as their data and tools, throughout their research:

> "Interpretation is at the centre of data analysis. Regardless of the size of a data set, it is subject to limitation and bias. Without those biases and limitations being understood and outlined, misinterpretation is the result. Data analysis is most effective when researchers take account of the complex methodological processes that underlie the analysis of that data." (boyd & Crawford, 2012, p. 668)

By the very uttering of a research question, a framing is produced that primes researchers to look for specific information. The question sets up patterns of relevance-recognition in the researcher. Research itself can be at least partly characterised as an extractive exercise ranging over a specified domain of interest—the construction of the research question itself is the first step in this extractive approach. Without careful attention a critical perspective can be lost and insight to the object of study can be obscured. The methodologies utilised in Big Data approaches in general are inductive in drawing on specific resources and constructing from them general outcomes. The rules of construction may be unclear to the researcher (e.g. because of the role of algorithms). Rather than the 'end' of theory, this seems to be more like a period of obscured theory. This has been described in terms of the 'selective focus' brought to bear upon patterns of data in neuroscience. Specifically with reference to fMRI studies. Jessey Wright describes the 'double-edged' role of data analysis and interpretation in this area:

> "...techniques render data interpretable, but their selection and application are often informed by the methodological and theoretical commitments of researchers using them." (Wright, 2018a, Chap. 11)

This serves to highlight an uncertainty that can be discerned in the role of data in neuroscience research. On the one hand, it appears to offer revolutionary empiricist material, while on the other, it appears to risk obscuring theoretical presuppositions and assumptions embedded within methods.

In her *Weapons of Math Destruction (2016)*, Cathy O'Neil highlights many of the problems that can come from feedback loops within algorithmic, data centred systems. These feedback loops tend to alter the contexts of which the data are supposed to be neutral pictures. In contexts of law enforcement, for example, they can produce an intensification of policing in areas already exhibiting high crime rates that in turn produces more arrests, and so more data on apparently increasing criminality. Data can serve to fuel the very rises it then reports on, leading to a self-reinforcing cycle of detriment to those on the receiving end, without understanding any of the possible explanations for why crime rates might be high in the first place. The principle of feedback loops here also emerges, especially when we think of the value choices made at the inception of brain data collection, and the subsequent use of data sets in framing ongoing research. With brain data—those widely aggregated, processed signals—we can expect to find what we're looking for because the search is framed by those data we seek. And we can expect to prize what we find because its salience is reinforced as it populates databases, further framing subsequent searches. Rather than an unbridled exploration of the objective brain, facilitated through data, the result may be a self-reinforcing dive into deeper detail on existing interests. To further posit cognitive and behavioural insights based on a potentially misunderstood endeavour would be not just a pity scientifically, but a ground for ethical issues in mischaracterising the nature and activity of human beings in a too-reductive way.

## What Brain Data Can Reveal

The nature of the order presented by datafication of the brain may not necessarily be one that is revealed as much as imposed. For sure, the brain is highly ordered, but the methods of data science already contain a characterisation of data, i.e. it is something that is transmitted from locations to other locations, via more or less complex networks. The brain fits this complex network model excellently and represents an ideal arena for the exercise of data science's most powerful techniques. But it isn't clear that because this is the nature of the brain it has straightforward implications for cognitive and behavioural phenomena. Because there is a neat symmetry between data scientific approaches and the functioning of the brain, there may be a sense in which this is taken to explain the cognitive and behavioural phenomena. But the models produced in data of these phenomena are still models and taking the model for the modelled is not methodologically warranted.

Data in neuroscience is, at least, a practical methodological necessity not just in the sense that any scientific observation is a datum, but more specifically in the sense that signal recording and processing plays such a big role in generating a detailed understanding of neural activity. But it can easily be taken as more besides. Might large-scale, processed brain data come to be seen as a basic level of explanation for cognitive and behavioural phenomena to which reduction ought to aim, just as the neuron doctrine has provided such a basis? The localisation problem might be re-cast in data terms, if so. The inscrutability of these data might be taken as implying their fundamental nature—the substrate in which other phenomenon beyond the neural are anchored. It's worth approaching this question by way of some examples of what brain data can reveal about a person. By doing this, it will become clearer just how important brain data can be in accounting for a person in a general sense. At the same time, it will become more apparent why important conceptual and ethical questions need to be addressed in terms of brain data.

What can data derived from brain signal processing reveal? The immediate thought might be that the contents of mind can be revealed. By recording brain activity and creating troves of data, could neuroscience start to read mental states, and make progress into revealing the stuff minds are made of? It might seem that, through experiment, recording of brain signals, and processing of signals with complicated algorithms and AI that the contents of minds become legible, besides the activity of brains. This, as we will see, is only partially true, and only in a very limited sense. Certainly, there is no reason to believe neuroscience can produce 'mind reading' as it is known from sci-fi or fantasy stories. What's worth making is a distinction between brain reading and mind reading, and concentrating on how data mediates the two. In that context, some of the most impressive neuroscientific brain reading experiments come from Jack Gallant and his research group. Gallant has used intricate experimental techniques to record brain signals from which have been decoded semantic information. A 'semantic map' of the brain has come from Gallant's work, which can predict the kinds of objects people are hearing about when exposed to natural speech. Based on brain recordings made of participants while they listened to a podcast, labels were created that tagged brain activity to the verbal input being undergone:

> "The labels assigned to the twelve categories were tactile (a cluster containing words such as "fingers"), visual (words such as "yellow"), numeric ("four"), locational ("stadium"), abstract ("natural"), temporal ("minute"), professional ("meetings"), violent ("lethal"), communal ("schools"), mental ("asleep"), emotional ("despised"), and social ("child")." (Huth et al., 2016)

This is a fascinating piece of research that gives insights to how semantic classification is represented in the physical brain. It certainly appears that some form of mapping of meaning can be derived under controlled circumstances from a person's brain activity. It looks like the opening of a way into the contents of the mind, at least in a limited sense. It's worth noting the difference nevertheless between 'semantic classification' and how meaning works in language. Semantic classification as a neural process may be automatic, a fact of neurophysiological and embodied experience drawn from the brute facts of human perception as evolved over thousands of years. The content of a meaningful experience one has when exposed to specific auditory stimulus may differ from the content of that stimulus. Hearing talk of a stadium may be classifiable in terms location, it might prompt a detectable neural response similar to any similar talk of locations of statues, schools, parks, but it may be experienced as emotionally contentful for a sports fan. Talk of a rival team's stadium location may evoke a negative emotional response, for example. Reading off the locational meaning from the brain might give clues as to the nature of the stimulus, but not so much the meaning of the experience. The latter draws upon interpersonal resources including culture, experience and learning, varieties of intention, context, in a way that the physical conditioning of perceptual capacities through evolutionary processes needn't.

The neural classification may give clues to what sort of meaning a stimulus has in some sense but may not account for it entirely. Similarly, the meaning of an experience may not have the expected relation to the stimulus, in much the same way that one can see something and completely ignore it. One might be looking intently at a monument to some historical event, while utterly absorbed in thinking about what to eat for lunch. The classification of the visual representation that can be reconstructed from brain data may have little bearing on what's significant to the person. It could have a variety of relations with conscious, intentional mental experience. Meaningfulness of auditory or visual perceptions depends on what the person is thinking, feeling, remembering, and so on at least as much as how one might classify the input stimulus. There is more at stake than the data might show.

Another, perhaps even more striking example of brain reading with implications for revealing the mind comes from Shinji Nishimoto, Gallant, and others' work on revealing mental imagery as derived from brain recordings made while research participants watch videos (Nishimoto et al., 2011).

When you look at a scene, you can see it to the extent that photons of light reflect from the surfaces of objects and collide with the interior surface of your eyes. The energy carried by those photons is transferred to your eye in different amounts according to where in the spectrum of light they stand. Colours tending toward blue bring most energy, those toward red, least. In each case of looking at a scene, this light energy is converted to a series of electrical impulses that are carried along the retina into the visual processing system of the brain. In the visual system of the brain, regions corresponding to the visual field are 'lit up' electrically, matching the spatial and energy distribution of photonic impacts from the eye. In looking at a scene, there is a mapping from light, through electrical activity, from objects in the world along to the brain's visual system.

If you look at the scene I'm looking at, you can usually predict what I'm seeing by imagining my perspective—*putting yourself in my shoes*. This echoes the sort of cognitive activity based on reasoning that was mentioned with respect to clinical radiographers and CheXNet. Putting yourself in the shoes of another revolves around connecting judgements via considered reasoning about perspective, place, and wider knowledge of the world. But what of the alternative approach also described there, dealing in probability and connections among likely-correlated inputs? Given the mapping between objects, retinal processing, and the electrical activity of the visual system, could you have looked 'inside' my brain and made a prediction about what I would report seeing?

In principle, if a clever neuroscientist were presented with a swathe of data derived from the visual system, they might be able to predict what was being seen by the person whose visual system they were examining. Small studies have been carried out that suggest this is technically possible, while noting there is much more work to be done. Nishimoto's work centred upon combining data from across the visual system of the brain in order to monitor energy peaks as videos were watched. From the energy variations over time, and the locations in the visual system of the brain that these variations ranged over, approximations of the contents of a participant's visual field could be reconstructed with 75% accuracy. While this work was on a small sample of just seven participants, it nonetheless is considered by the researchers themselves to be "…a critical step toward the creation of brain reading devices that can reconstruct dynamic perceptual experiences." (Nishimoto et al., 2011, p. 1644)

Instead of imagining the perspective of another person, and predicting the objects they are looking at, it is *possible* to monitor just the brain and predict visual stimulation based on electrical activity. From this, the objects acting as stimulus can be reconstructed. Reading the brain can provide insight to what a person is *seeing*. Reading information from the visual system is not restricted to what is seen while awake. In dreams, the visual system shows activity similar to that when stimulated by objects seen in the world. The difference here is that the system is responding to dreamt images. These are truly mental images, without objective, real-world counterparts. And indeed, experimental work has shown some success in decoding dreamt images along the same lines as those from awake subjects (Horikawa et al., 2013). In studying dreamt images, brain recordings are combined with images of activity levels across the brain, and testimony from those having the dreams (once they are awakened). From this combination, some of the contents from dreams can be correlated with brain activity. This provides a basis for predictions about the contents of dreams.

By recording electrical activity in the brain, it is possible to predict what a person is seeing, and the images conjured in her dreams. On the one hand, this is fascinating science from which a great deal of insight might be drawn. On the other, it may cause some unease in that the hitherto private domain of the mind might appear to be somewhat open to view. And this apparent openness is not limited to visual phenomena. Through decoding electrical activity in other regions of the brain, quite fine-grained information may be sought from throughout the auditory, cognitive, memory, mood, and other systems of the brain. In addition, as data is increasingly collected, the power of algorithms grows. Algorithms depend for their power upon having been 'trained' on large databases. The larger the databases, the better trained the algorithms. As a turn toward data begins, the more material there is for algorithms to be trained on. This in turn leads to more productive data-centred neuroscience, and a cycle reinforces itself.

With more data, and more powerful algorithms, some expect neuroscientific discovery to extend further into what have been hitherto considered quite abstract dimensions of mental life. These include responsibility, or the capacity for responsible action, based on neuroimaging analysis rather than established factors like age (for criminal liability, for example (Farahany, 2011)). Others imply that the coarseness of brain function categorisation could ultimately be tuned-up through the judicious use of neuroimaging, thereby boosting our understanding of mental disease and brain function through data (Bzdok & Meyer-Lindenberg, 2018). In the split brain experiments, researchers were confronted with outward behaviour that they could account for better than experimental participants. This was based on knowledge about the specifics of the research set up. Here, researchers are presented with troves of data which seem able to get beyond behaviour. This seems to be what drives the interpreter hypothesised as active in those experiments, rather than its products. Looking at the signals recorded from brains, processed by algorithms, researchers seem to get at the heart of what produces cognition and behaviour.

Gallant's group are very clear that the data-driven nature of their experimental work is crucial to its productivity. In order to produce semantic maps of the brain, an algorithm was developed that allowed information gathered from small groups of research participants to be scaled beyond specific results in order to create a model. Doors to such modelling based on small samples are opened by a concentration on data. While personalisation requires subject-specific information, and in specific research contexts this personalisation is required for robust results, the overall data-driven model provides a powerful and generalisable template. As we come to consider consumer devices the role of algorithms will become even more central, as they allow this data-driven mode of operation that can go beyond sparse data to make far-reaching predictions. It is characteristic of a discovery science approach through data, as opposed to the traditionally dominant mode of hypothesis-driven scientific investigation—a *see what we can see* approach. With such open-ended practice, though, there is a risk where sufficient controls are lacking.

The example of the visual system, and reference to responsibility-gauging, shows in principle how more wide-ranging predictions could be forthcoming in practices based in brain data. These could include how brain data can relate brain states to phenomenal states, or to dispositions. The data-centred approach allows for powerful advances beyond the specific, to the generalisable, especially through the development of algorithms. This was evident most clearly in the step described in Gallant's work where general brain activity models are developed from specific brain data. This suggests the ways in which data can be leveraged to create models more than the sum of their parts. Through the convergence of data processing techniques applied throughout the brain, by way of a fusion of approaches and with a variety of aims, even more scope is opened not just to read the brain but to manipulate it too.

Whole brain computerised modelling aims to provide a dynamic picture of brain states by combining data from different sources (such as electrical activity, structural information, BOLD signal) and relate them to provide a picture of the brain in action. This is 'brain activity' on a different scale to the neuronal-reduction picture inaugurated in Ramón y Cajal's vision. In being a relational picture of a variety of signals, this approach produces images that really portray series of relations among complex data. The outputs here are based on 'data-data relations' because unlike the production of an fMRI image as based on a table of data, this approach takes varieties of tabulated data from across brain areas and of varying types, and relates those tabulations to one another. These are input to a computer system based on the intrinsic and complex activity of a whole-brain model, which can then be used to exhaustively model all available permutations of brain state dynamics latent in the data. The dynamics of whole-brain computerised modelling are *data dynamics*. This is from where the power of the approach derives, in fact, as it is in a certain sense agnostic about details of which neuron fired when (an important element for EEG signal processing) but instead takes a panoramic view of brain activity as can be derived from the data. It is a data first approach, in this sense. In work by Morten Kringelbach and Gustavo Deco, for example, 'brain state' is redefined from a static time slice of neuronal firings at a time to a 'probabilistic cloud' of transitional states based in swathes of brain data (Kringelbach et al., 2020).

In their work, some of which focusses on prompting transitions between brain states, Kringelbach and Deco present "...a mechanistic framework for characterizing brain states in terms of the underlying causal mechanisms and dynamical complexity." (Kringelbach & Deco, 2020, p. 1) In practice, this means brain states seen as an ensemble of neuroimaging, spatio-temporal dynamic, empirical behavioural, EEG, Magnetoencephalography (MEG), fMRI, and PET data. This ensemble presents a picture of brain states that includes their dynamic relations to one another. This heterogenous approach allows whole brain modelling *in silico* that aims to strike a balance between complexity and realism to provide a realistic, yet still computationally workable, model of the living, human brain *in vivo*. In principle, this approach would put in the groundwork to permit instrumental intervention upon the brain in order to prompt transitions from one brain state to another:

> "The goal of whole-brain modelling is to predict potential perturbations that can change the dynamical landscape of a source brain state such that the brain will self-organize into a desired target brain state." (Kringelbach et al., 2020, p. 5)

If perturbations can be predicted, modelled in silico, then they might also be created as stimuli that can serve to bring about self-organisation into a target brain state. One such transition discussed in the literature is that between waking and sleep states.

Deco and his group used stimulation of whole brain models in silico to examine induced transitions between model sleep and wakefulness states, with some success in prompting changes from one to the other (less with wakefulness to sleep). This kind of technology is in its infancy. But basic scientific payoffs are hotly anticipated, as are the eventual hoped-for therapeutic rewards:

> "The approach could be used as a principled way to rebalance human brain activity in health and disease[] by causally changing the cloud of metastable substates from that found in disease to the health in order to promote a profound reconfiguration of the dynamical landscape necessary for recovery, rather than having to identify the original point of insult and its repair." (Deco et al., 2019, p. 18094)

The predictions made about whole-brain dynamics are based on how patterns of data inter-relate across the brain under a variety of circumstances. For instance, researchers have observed subjects as they fall asleep from a waking state. Using arrays of different types of data, wakeful states, sleeping states, and the transitional dynamics from one to the other can be modelled. Rather than a static time slice, this dynamic picture portrays tendencies within interacting systems across the whole brain that map states and the coordinated changes among them. One advantage of this technique comes in being able to model transitional states. Transitional states are particularly interesting because they remind us that the brain is never *in*active, and that the isolation of a 'state' for research purposes is an abstraction of that set of parameters from an array of others. Moreover, in understanding transitional states, researchers gain some understanding about what prompts the complex set of brain activity present in any specified state to shift to another.

To borrow an analogy from meteorology: we can understand rain and dry weather. This is like having information on individual brain states. But once we understand a specific phenomenon like barometric pressure and how, when it falls at a certain rate, rain results we have a powerful predictive tool. Moreover, if we had the technology to prompt widespread barometric pressure changes, then under favourable conditions maybe we could make it rain by utilising this knowledge. In the case of the brain, understanding transitional states is like understanding barometric pressure. Given what's already known about how the brain works, we already have tools that can alter brain dynamics. So with growing understanding of brain state dynamics and the transitions among brain states, we can move toward the ability to manipulate brain state transitions. In the future, for example, with the dynamics of wakefulness and sleep transitions understood neuroscience could potentially produce technologies capable of prompting transitions between those states. Such instrumentalisation of the brain could permit drug free anaesthetic, or potential treatments for coma patients, as brain stimulation might be put to use in prompting sleep from a wakeful state or prompting wakefulness from sleep states. Imagine:

> It's 40 years from now. Ada is going into hospital for routine surgery. In the past, general anaesthetic was administered through a cocktail of drugs. This was an inherently risky approach, with careful monitoring required. Breathing especially had to be closely observed. After general anaesthetic of this kind, moreover, cognitive deficits could appear and last for an extended period, post-surgery. Now, Ada need only wear an electromagnetically active headset while algorithmically-optimised electromagnetic fields inhibit and excite specific brain areas so as to switch her brain from a wakeful, to a deep sleep state. This state is maintained dynamically by the same system throughout surgery. Upon completion, she is awakened once more by electromagnetic stimulation, benefitting from drug and side-effect free anaesthesia.

The kind of instrumental control over brain states in this imagined vignette would be one beneficial outcome from investigations of brain data dynamics. Where brain data dynamics become the object of research themselves, manipulation of them becomes an obvious focal point. If transitions among brain states can be controlled, and instrumentalised, the effects brought about by physical interventions on those brain states could in principle be used to bring about target *mental states*. Beyond the idea of drug-free anaesthesia, applications in this arena could target psychiatric problems where those problems are identified as neural problems. Brain states associated with depression or anxiety, for example, might be targeted for transitions to brain states more associated with calmness or clarity of mind. In terms of human enhancement too there could be applications with differing levels of ethical concern—enhancing a surgeon's wakefulness for long surgeries, for instance, versus that of combat soldiers, or bomber pilots, to optimise battlefield readiness. While such outcomes remain remote for now, they are among the end points envisioned for a mature brain-state transition paradigm.

Those such as Kringelbach working in this area acknowledge the central role of data in their work. Without algorithmic processing of swathes of brain data, these approaches would not be possible. As with reading the brain to map semantic classification, or to reveal mental imagery, the focus on data allows the development of techniques that go beyond specific results and that can ground generalisable approaches.

Among philosophers, there can be found a generous helping of skepticism about much of the potential to read off mental states from brain states. Much of this is well-grounded. For example, Valerie Hardcastle and Matthew Stewart offer a biting analysis of unstated assumptions in neuroscientific research that would seem to limit its applicability. They are especially critical of the 'Subtraction method,' used in experimentation. This involves picking pairs of related brain processes assumed to co-vary with respect to a perception or cognitive activity under investigation, and then measuring differences in activity in brain regions associated with those processes. From examining variations in the regions of interest, neuroscientists come to believe they find relationships between neural and cognitive processes. Hardcastle and Stewart suggest that this is erroneous, as it is based in unstated assumptions:

> "Neuroscientists' simplifying assumptions of discreteness and constancy of function are simply not justified, nor is it clear that they will be theoretically appropriate any time soon. These assumptions end up causing scientists to abstract over the neurophysiology improperly in that they deny diversity of function or multi-tasking all the way down the line prior to examining the data for exactly these possibilities." (Hardcastle & Stewart, 2002, p. 9)

Brain data as it is being discussed in this book represents a general aggregation of many neuroscientific findings. If Hardcastle and Stewart are correct in their take on basic misapprehensions within the disciplines of neuroscientific experimentation, then brain data can mislead. It is too far from the phenomena of the brain, never mind the mind, and too tinged by presupposition. If basic neuroscientific research is flawed regarding uses of the subtraction method, what hope could there be for taking research results and combining them wholesale? While the subtraction method of brain data analysis might indeed require careful scrutiny, Jessey Wright offers a broader analysis of neuroscientific research techniques including an array of other approaches. Wright's point is that neuroscientists are aware of the limitations present in their approaches and analysis techniques, and so they use multiple techniques to boost the robustness of their experimental analysis:

> "The general lesson of the experimenter's regress is that problematic assumptions can arise in the context of experimentation. The general lesson of the appeals to robustness is that those assumptions can (sometimes) be validated by comparing different perspectives on the same subject... Thus, the debate about the epistemic status of neuroimaging, which is framed in terms of the logic of subtraction, is at best an evaluation of the limitations of analysis techniques that depend upon that logic. Sweeping conclusions about the range of hypotheses that neuroimaging technology can and cannot be used to investigate are not supported by this literature" (Wright, 2018b, pp. 1199–1200)

On a case by case, experiment by experiment, basis where data can be shown to be modulated by specific experimental manipulation, robust findings can be possible through careful attention to detail. There may remain nonetheless a further issue when data are coalesced into vast datasets, themselves then processed in global fashion using techniques like that of 'compressive big data analytics,' or CBDA, discussed earlier in terms of how huge datasets could be automatically sampled without guiding theory to produce things *we didn't know we didn't know*. How can robustness and independence be sought under these conditions, where access to the nature of processing, and the intention behind data collection, is not available. In these contexts, we are not talking about relations among experimental design, signal, data and results anymore, but data to data relations guided by algorithmic operations in a very general *ad hoc* sense. Hardcastle and Stewart highlight problems in assumptions about discrete functions residing in specific brain regions. Wright suggests mixed-methods and explicit attention to detail can nevertheless ground compelling results. But the data-to-data context may provide novel issues orbiting around this nexus of concerns. The idea that answers to as-yet unasked questions reside in vast data sets is an unstated assumption that cannot really be well tested or mitigated by careful attention to methods and research design. The assumption is just a part of the data aggregation approach, a part of the rationale of 'big data' like processing, but part of the same approach effectively obscures—in Wright's words— *the methodological and theoretical commitments of researchers using them.*

What's key is that the kinds of insights here discussed are possible only through data, and its processing by powerful algorithms. The results of this processing seem to reveal the states of the brain that are relevant to specific instances of cognition and behaviour. They can predict perceptual contents, attention to those contents, and types of abstract disposition. In underlying behaviour, in being causally prior to any outward behaviour, these data-derived predictions from the brain bring with them an air of basic truth. Unlike the split brain experiment, it seems as if here behaviour needn't be interpreted and compared to the parameters and aims of an experiment. Rather, through complex processing, it seems as if the facts of the (grey) matter are laid bare. Certainly, this would need to be the case where brain state dynamics would be manipulated to prompt transitions among those states, to alter cognitive and behavioural performance. But any such instrumental interventions would demand scrutiny in terms of the data. What values are at work in terms of data recording, database curation, transparency of processing? These likely unstated values provide the basis for an array of novel ethics considerations, based in this datafication of brains.

# Ethics Concerns: Dashboarding

A brain data revolution in neuroscientific research promises a means of understanding cognition and behaviour in unprecedented ways, through data processing. It also includes the potential risk of reducing persons to their brain activity, and prioritising statistical models of neurobiological subsystems over the person's broader interests. Neuroscience appears to give unvarnished access to the brain-based correlates of cognition and behaviour, through detailed brain signal recording. This seems to shed light on the processes that underlie conscious activity, by allowing researchers to peer into the inner working of the brain. Computer processing complements this with a dynamic inferential capability, processing the brain recordings and scrying patterns latent within them. There is wide scope for re-imagining human nature on these grounds, as if brain data were the ultimate basis from which to account for thought and action in general. But there are important issues at stake in this datafication move. These occur in epistemologically and methodologically obscure ways.

Where CBDA is at work, for instance, and 'the end of theory' is assumed, it isn't clear how we ought to consider the knowledge that emanates from the processing of data. How ought we to think about what we know after algorithms present us with patterns in data? We do know that data are the result of recordings from brains, and that algorithms trained on datasets pick out patterns in that data. But the results may be taken for objective truths—patterns really inherent in brain activity. This is not what can be yielded by processing like CBDA, however. Such processing can only present its outputs relative to the material present—to the data set it has to work with, as complex and multifaceted as that might be. The output will always be a snapshot from selected sources. What's more, from instance to instance of processing, it is far from certain that 'the same' methods are being used. The overall, general methods of data analysis are describable in the same terms. But the specific application of learning algorithms in particular will be influenced by the data on which those algorithms are used from one instance to the next.

As data changes, grows, is cleaned, pared down, or otherwise changed, the function of the algorithm will alter with it. Maybe this is barely perceptible in terms of analysis from input to output, but any change would be real. In some instances, to avoid an algorithm being too rigidly constrained by its training data, elements of randomness are introduced to its functioning. In a strict sense, it's hard to say of some algorithms that the same one is used more than once. It may strictly perform a slightly different function from one instance to the next, owing to learning from intervening uses, or to randomness introduced to boost usefulness. 'Using an algorithm', in short, is not like using another more conventional physical tool. What can be done with such tools is limited by rigidly definable properties on which expectations can be built. Algorithms can ground expectations, but how they lead to their outcomes may not be known and may not proceed via the same means from one instance to the next.

Looking at how data is gathered will serve to crystallise the issue here. In a generic example case of using EEG recording of brain activity in response to some stimuli, the experimental set up is framed by a research question and informed by data about typical behaviours and neuroscientific knowledge. This kind of knowledge, among other things, informs where electrodes will be placed on the head. An experiment looking into the neural correlates of speech, for instance, may place electrodes on the left side of the head in response to the left side of the brain being dominant in verbal contexts. The signals recorded will go on to be processed by algorithms chosen for their established usefulness in the kind of experiments being undertaken. Neuroscientists interested in speech will think of a different set of algorithms compared to, say, others interested primarily in motor activity or vision. With signals processed, the data retrieved from them will be databased. The database will itself then serve to inform and optimise future uses of those algorithms.

There are value choices made in terms of framing a research question, in selecting recording sites and methods, and in selecting appropriate algorithms for the experimental case. By way of example, these value choices will include 'framing of questions' because the pursuit of specific questions might rely on obtaining research funding. Depending on prevailing research funding regimes and kinds of research sought, different aims and types of approach will be favoured as a result. For instance, where research funding agencies prefer technology development as an endpoint for research, researchers would be required to explain how their basic research enabled future technologies before they gained funds. This would have the effect of emphasising research that could more easily demonstrate technological applications, while relatively side-lining basic research aimed at non-applied ends. Relatedly, this would go on to condition at least to some extent the choices made about which equipment to use.

Researchers might prefer specific hardware or software, for reasons of future technological interoperability or commercialisation. This would have knock-on effects for what was considered a 'good' or 'good enough' signal acquisition, with further effects on how signals could be cleaned, processed, aggregated, and databased. Moreover, the data retrieved in specific experimental cases, once databased, goes on to characterise algorithm operation in the future by informing and optimising them. That data becomes the learning set for subsequent algorithms. What is discovered through experiment is thus connected to the research framing by way of justification (see Fig. 1).

If I predict that falling barometric pressure leads to rain, I use a context of falling barometric pressure and hope to discover the beginning of rain. Where I do indeed find pressure dropping and rain resulting, I then justify my prediction by pointing to the rain. I justify my discovery in light of the outcome. But in the case of data and the brain, data appears on both sides of the equation. The databases feature in the material that creates research questions, and that contributes to framing experimental set ups. What gaps in knowledge there are will be gaps among the collected data. Where, in detail, I ought to look in the brain to address the knowledge gap will be informed by the data derived from earlier experiment in a kind of confirmatory loop between discovery and justification. Subsequent interpretation of the data retrieved ought to take this into account.

This is somewhat the form of inductive reasoning in general, or of empiricism in general. But in this case, there are obscurities which make for problems. Brain recording and processing by algorithm, appears to be deriving objective truths from the brain. Overall, the promise of data appears to be the disclosing of ultimate neural truths that serve to explain

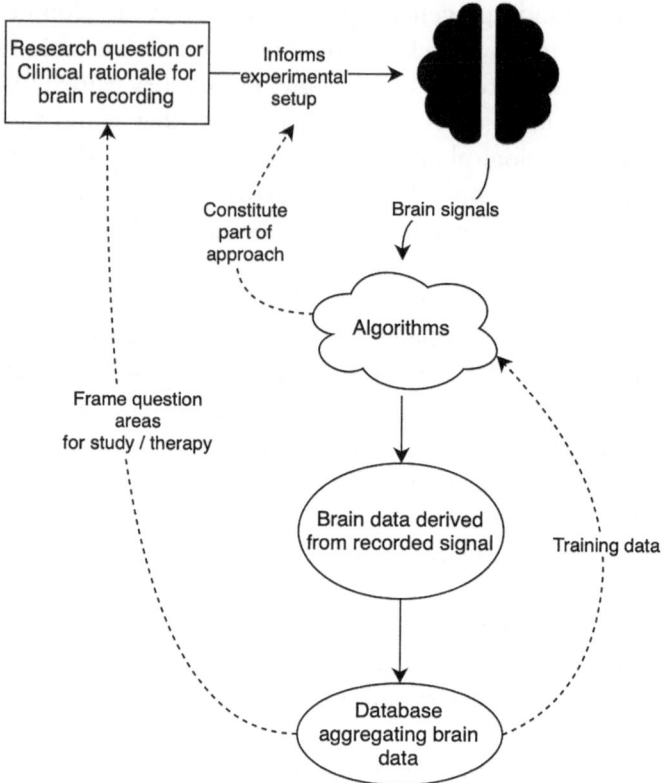

**Fig. 1** A simplified depiction of a brain data cycle: Algorithmic processing of the recorded brain signal can produce outputs relevant for research, for controlling hardware and software, and for clinical applications. But in furnishing databases too, it produces a feedback loop wherein the same data serves to train and optimise existing algorithms. These then go on to produce subsequent data, through processing brain signal recordings, and reproducing the cycle. Meanwhile, research findings and clinical data will go on to inform future generations of exploratory brain research, leading to brain signal recordings further related to the operation of algorithms. The data-to-data cycle reinforces itself through algorithmic training as well as through steering research and clinical findings

cognition and behaviour in ways not even available to the person. But in framing research questions and serving to justify the outcomes from experiments so framed, this confirmatory loop looks more like it is constructing a self-reinforcing picture of a datafied brain. Immutable facts of nature are not being deduced from brain activity. Explanations are being constructed from the processing of data derived from brain activity, with respect to specific circumstances, via algorithmic processes. This is not to say the findings from this kind of approach are thereby invalidated. But it is important to know what we are getting from science if we are to understand it and use it correctly. Perhaps especially when it comes to the brain, there are high stakes in this correctness for individuals.

It is very easy to buy into personal identity in terms of *brainhood* (Rose & Abi-Rached, 2014). Given what we know of phenomena like that illustrated in the split brain experiment, and about mismatches between intention, will, and self-deception about our own behaviours, it could appear that we are prone to varieties of ignorance about ourselves. Science in general is something widely thought of as aiming at ultimate truths. Why not have a science of the brain as a science of personhood? After all, the brain is at the centre of all we think and do. If we know how it works, we can know how we work. And if we can control the brain, we can gain ultimate self-mastery, free from self-deception and weak will. Perhaps especially with respect to whole-brain computer modelling approaches to the brain, and the associated prospect of forcing brain state transitions, the idea of *data dashboarding* can be seen to arise.

A dashboard in a car often simply displays information about speed, engine state, maintenance requirements and so on. In software contexts, a dashboard is a dynamic tool that both indicates statuses and permits intervention. If a laptop is slow, for instance, a user can open a dashboard and see what's happening. If many programmes are running, the user may suspect this is causing overall system slowness. From the dashboard, they can stop a few applications and evaluate any performance improvements. It is a small step to take to see brain data as reporting brain states to being adjustable parameters for those brain states. As with the aspiration to prompt states of wakefulness to sleep and vice versa, so to for any state captured in a whole brain model.

If we could have a dashboard that offers the chance for dynamic intervention upon our own brain states, reified through recorded data, how ought we to understand the use of that dashboard?

The datafication of how we understand the brain, coupled with the potential for whole-brain computational modelling opens the possibility for re-imagining the brain as a kind of control panel for moods, dispositions, beliefs, and ultimately *ourselves*. We don't need to decide whether the prospect of such literal self-control is good or bad. But the very idea requires conceptual and ethical scrutiny. For one thing, this would represent a change in the idea of persons as generally understood. In terms of autonomy, or the ability one has to decide for oneself about what to think and do, would self-control in this instrumental sense register as a boost? On the one hand, it seems yes. One might be newly enabled to decide to be another version of oneself, with greater ability to enact the sorts of things hitherto out of reach. On the other hand, we might ask whether the person who decides to switch brain states for the first time is the very same person as the person who results.

We could imagine a scenario where a shy and anxious person wishes to be more confident in their day-to-day life. That person could seek out a device to manipulate their brain state in order to realise less inhibition. Perhaps such a device could be found, and perhaps it could work, in the sense that the device user's shyness diminished. To what degree could shyness be diminished before a change in that person might be considered more than just a dispositional adjustment? What if disinhibition came to manifest arrogance, or recklessness to the extent that friends and family began thinking of this once shy person as transformed?

While the device user might be happy to remain under the influence of the device, having tried and failed to be more confident in their life until now, if they were to become distinctively changed it might be said that they are not a more controlled version of themselves, but a different self. In evaluating the change, those friends and family might think that this isn't their newly confident friend or relative anymore. Rather than confidence, as that word is normally understood, this is to them a shy person acting out of character owing to a brain intervention. Would they be wrong to think this? Would the previously shy person be justified in thinking themselves newly confident? If their shyness was reversed to the extent of recklessness, it might be the case that we would question their ability to recognise if a device was going too far in affecting their disposition. They would, after all, *be newly reckless* and so from that position presumably care less about such effects.

Psychological traits or dimensions of personality are typically thought of as related to the varieties of experience and context a person enjoys, endures, or otherwise undergoes. There is something organic and self-making about the nature of this experience that gives some content to those traits and dimensions. People are shaped by experience, including how they react to it. Given the data-centred, brain state switching nature of transformations from, say, an anxious to a confident person, it seems a thinner sort of 'experience'. We might even want to withhold the ascription of the term 'confident' from the latter person, instead suggesting something like *disinhibited*. Here the role of language and meaning comes into contrast with the instrumentality of data driven dashboarding.

'Confident' and 'disinhibited' mean different things. But in an imagined scenario where one uses a dashboard to boost one's 'confidence,' the intervention might just as well be labelled disinhibition by neuroelectrical dynamic means. This means different things, and implies different things, for the newly dashboard-modified person. By the very fact that the dashboarding intervention is something new, it lacks the historical and cultural dimensions that the word 'confident' has. Confidence might be more or less favourable as a disposition among different people, among different context, times, and circumstances. The very word may connote different things in different languages. But a dashboard-driven intervention would be a newly minted practice in need of evaluation in light of the ways in which it was used. It would require contextual appraisal in order to establish meanings for it. It is not enough to simply posit it as co-extensive in meaning with 'confidence'. Whether and how the new term might become co-extensive, or how it might interact with established meanings of words like 'confident' is an open question. But it can't be defined into one significance over another.

Importantly, what this illustrates is that the simplifying tendency of datafication in brain signal recording and processing might miss out the phenomenology of human experience. We can recognise the difference between being confident and being disinhibited, especially from the first-person point of view. But it is not clear that the datafication process that underwrites brain state switching represents that difference. Imagine an example case:

Cara is affronted at having been excluded from speaking at an important meeting because a prior speaker has gone over time. It is an affront and a *de facto* dismissal of her ideas. Dominic, the member who has filibustered, is affronted that Cara should dare question his right to occupy all the time he sees fit. He thinks Cara should learn to *play the game* better.

Cara and Dominic are equally affronted. But Cara has significantly different reasons from Dominic for justifying her being affronted. Indeed, whether what each counts as being 'affronted' really is *in fact* might be up for debate, regardless of their genuine feeling. 'Being affronted' is not a neutral condition of the person, but one that is evaluable in terms of its reasonableness or not. The reasons offered in justifying the state will serve to characterise that very state. The whole-brain state corresponding with 'being affronted' however (assuming one could be found), would be a generic sort of state. This would be an a-rational one, perhaps, or one based in reasons not related to a specific person and the course of their experience. The content of this state would represent the effect of a neuro-stimulatory cause. It might well issue in behaviour we recognise as that of being affronted. But again, we may be tempted to refrain from even calling this real affront at all, as such states require a richer contextualising story than the presence of a specific brain state.

The dispute between Cara and Dominic about who really ought to be affronted, or whether they were really something else, is different from this. There, each party's *reasons* are at stake. Here, there are not the same kinds of reasons. There are instead observations, aggregations of data, and correlations among cognitive and behavioural outputs. Accounting for brains in terms of data might be seen as casting perspectivally rich points of view as rather generic. Or, as casting individual persons as generic experience-havers, whose reactions are apt for adjustment according to patterns latent within data derived from neuroscientific research. The question isn't whether the dashboard-driven brain-switching approach 'works' or not—by hypothesis, it does indeed alter brain states and thereby produce cognitive and behavioural changes—but whether or not it works on the right basis.

Typically, when people adopt a point of view, or make decisions, they at least believe themselves to be deciding among options for good or bad reasons. They don't typically think of themselves as merely realising brain states according to emergent norms implicit within data sets. The brain data account, with its probabilistic cloud and potential for data dash-boarding, would appear to provide a means of explaining away some influential views of agency in this way. The account of agency most salient here involves the capacity for someone to deliberate about what to do, or to have practical decisions to make about problems.

The kinds of agency boost afforded by a rehabilitative intervention, like a prosthetic limb, is agency boosting in terms of effectiveness of agency. The case of brain data is different. This is one way of stating a general problem of determinism, with respect to the brain. If the brain is the hub of human thought and action, then unless it operates primarily in a non-linear fashion, full knowledge of its working ought to allow for total prediction of every thought and action. For the fully clued-up neuroscientist reading the various signals of their own brain, there could be no deliberation about what to do, say, think, because they would know by examining their brain data readouts what they were about to do, say, and think. They could predict perfectly their thoughts and actions before they happen, and, in having no place for deliberation, could never have any practical problems. Everything would just *be*, naturally. No other path would be possible, either metaphysically, or epistemically.

But there's a conceptual muddle here. For one thing, with people, observations affect the nature of the object being observed. Let's imagine Ellen, a neuroscientist examining her own brain data:

> Ellen is observing her brain data. As she does so, the data change as a result of her having observed them. Now that she notices this change, the data change again. Every state is coupled with every perception and judgement. By the time Ellen begins to formulate a prediction about her brain state, based in her brain data, they change again as a result of these acts.

In this dynamic picture of things, there may be room for wondering about what comes next; to deliberate about the nature or the magnitude of the changes induced by the observation of oneself by oneself. But this might not be enough to make a lot of sense out of Ellen as being capable of deliberating, and thereby having practical problems. It is also a bit strange, because it seems to place the possibility for fragments of self-ignorance at the very core of agency. The salient account of agency here being used is centred on deliberation and decision-making about problems. But through Ellen's self-observations via data, she can only really be thought of as an agent to the extent that she doesn't know yet what she might do next. Ellen can only deliberate about her own states based in what she *doesn't* gain from the data. It seems odd to say, 'Ellen is free because she can't predict what her brain's next state may be.' It points to incompleteness in data as providing an account of agency.

There is another closely related conceptual muddle here. If we stay with the current account of agency, it is bound up with freedom as well as deliberation and decision. Having the scope for deliberation and decision-making about problems is an incomplete story without any capacity to act. It would be a thin account of agency that stopped before the follow-through to action. The role of deliberation is to deliberate among reasons to act one way or another, or to choose one action over another. Implicit in this is that decisions are to be made where practical problems are thought to genuinely arise. The problem part is *what to do*, so the issue is a decision. But in the kind of story just outlined, in which Ellen is observing her neural signals and deriving predictive power over herself, 'decision' is redundant. If she knows one option is closed to her—refraining from doing x—she cannot consider her options. If the data, the facts, the science, really show that she *will do x* then the decision to do x is redundant (For more wide-ranging analysis of this point, see Caruso, 2021). The main point here is that one's understanding of oneself as an agent is negatively affected by a purported form of deliberation over foregone conclusions. There is no agency-affirming way to deliberate about deciding to do x if it is known that there is no option open but that x will come about. Compare these descriptions:

A. Ellen is observing her brain data. She sees from the data that she is about to raise her arm.
B. Ellen is observing her brain data. She sees from the data that she is about to decide to raise her arm.

In (A) the data may show characteristic electrical activity in the parts of Ellen's brain that control her arm muscles. So she knows, all things being equal, her arm will rise. But (B) is muddled. Ellen can't *predict* that she will decide. To decide can't be predicted, on pain of it not being a decision but a determination. Jenann Ismael is useful here as she discusses control and choice:

> "There is no simple, general relationship between environmental stimulus and behavior, no fixed structure that can be exploited like the levers on a toaster to bring them under our control. You can make frog tongues flick and change the direction of a school of fish by producing the right stimulus. You can make cells secrete and dogs salivate and flowers bloom by producing the right stimulus. But it is very hard to control the voluntary behavior of another human being, because choice effectively randomizes the connection between stimulus and response." (2016, p. 96)

The only way Ellen can predict everything she is going to do is to have a situation like in (A), and have no decision-making involved. Her behaviour just stems from her neural states. What's at stake here is how we ought to think of brain data: do the data explain the corresponding cognitive-behavioural states, or *explain them away*?

These kinds of questions become acute where brain data might be employed in psychiatric accounting for human behaviour. Here, questions of whether thoughts and behaviours are 'normal', variant, or pathological arise. And coupled with an increasingly neuroscientifically-informed practice of psychiatry, brain data appear once more here. In trying to boost, or optimise agency through data dashboarding, we might end up erasing it. Instrumentalisation of discoveries based in a datafied approach, moreover, might create questions about the kind of 'self-control' afforded in a context of dashboarding. With such uncertainties, it is clear such brain data ought to attract specific and dedicated ethical reflection in terms of both their conceptualisation, and their uses.

An acutely ethically significant use of brain data comes in psychiatry. Here, human cognition and behaviour are assessed in terms of their normality, deviation, or pathology as regards a given set of normal parameters. Data could serve a revolutionary role in psychiatry, not least owing to its potential scale of application. But given questions around the meaning and legitimacy of its uses, a highly datafied psychiatry will require close analysis. Massive aggregation of data may well permit novel insights: but not necessarily more acute clarity. It will be a novel, perhaps highly predictive, view of the brain. But unlike specific experiments in which data can be carefully manipulated and used in quite rigorous and scientifically valid ways, huge aggregation gains its power despite losing track of the minutiae of specific data curation, collection, and validation.

## Bibliography

Anderson, C. (2008) 'The end of theory: The data deluge makes the scientific method obsolete', Wired, 16(07).

Bashashati, A. et al. (2007) 'A survey of signal processing algorithms in brain–computer interfaces based on electrical brain signals', Journal of Neural Engineering, 4(2), p. R32. doi:https://doi.org/10.1088/1741-2560/4/2/R03.

boyd, Danah and Crawford, K. (2012) 'Critical questions for big data', Information, Communication & Society, 15(5), pp. 662–679. doi:https://doi.org/10.1080/1369118X.2012.678878.

Bzdok, D. and Meyer-Lindenberg, A. (2018) 'Machine learning for precision psychiatry: Opportunities and challenges', Biological Psychiatry: Cognitive Neuroscience and Neuroimaging, 3(3), pp. 223–230.

Caruso, G.D. (2021) 'On the compatibility of rational deliberation and determinism: Why deterministic manipulation is not a counterexample', The Philosophical Quarterly, 71(3), pp. 524–543. doi:https://doi.org/10.1093/pq/pqaa061.

Cheour, M. et al. (2001) 'Mismatch negativity and late discriminative negativity in investigating speech perception and learning in children and infants', Audiology and Neurotology, 6(1), pp. 2–11. doi:https://doi.org/10.1159/000046804.

Choudhury, S. and Slaby, J. (2016) Critical neuroscience: A handbook of the social and cultural contexts of neuroscience. Wiley.

Deco, G. *et al.* (2019) 'Awakening: Predicting external stimulation to force transitions between different brain states', Proceedings of the National Academy of Sciences, 116(36), pp. 18088–18097. doi:https://doi.org/10.1073/pnas.1905534116.

Farahany, N.A. (2011) 'A neurological foundation for freedom', Stanford Technology Law Review.

Hardcastle, V.G. and Stewart, C.M. (2002) 'What do brain data really show?', Philosophy of Science, 69(S3), pp. S72–S82. doi:https://doi.org/10.1086/341769.

Horikawa, T. *et al.* (2013) 'Neural decoding of visual imagery during sleep', Science, 340(6132), pp. 639–642. doi:https://doi.org/10.1126/science.1234330.

Huth, A.G. *et al.* (2016) 'Natural speech reveals the semantic maps that tile human cerebral cortex', Nature, 532(7600), pp. 453–458. doi:https://doi.org/10.1038/nature17637.

Kitchin, R. (2014) 'Big data, new epistemologies and paradigm shifts', Big Data & Society, 1(1), p. 2053951714528481. doi:https://doi.org/10.1177/2053951714528481.

Kringelbach, M.L. *et al.* (2020) 'Dynamic coupling of whole-brain neuronal and neurotransmitter systems', Proceedings of the National Academy of Sciences, 117(17), pp. 9566–9576. doi:https://doi.org/10.1073/pnas.1921475117.

Kringelbach, M.L. and Deco, G. (2020) 'Brain states and transitions: Insights from computational neuroscience', Cell Reports, 32(10). doi:https://doi.org/10.1016/j.celrep.2020.108128.

Lotte, F. *et al.* (2018) 'A review of classification algorithms for EEG-based brain–computer interfaces: A 10 year update', Journal of Neural Engineering, 15(3), p. 031005. doi:https://doi.org/10.1088/1741-2552/aab2f2.

Marino, S. *et al.* (2018) 'Controlled feature selection and compressive big data analytics: Applications to biomedical and health studies', PLoS One, 13(8), p. e0202674. doi:https://doi.org/10.1371/journal.pone.0202674.

Nishimoto, S. *et al.* (2011) 'Reconstructing visual experiences from brain activity evoked by natural movies', Current Biology, 21(19), pp. 1641–1646.

Rose, N. and Abi-Rached, J. (2014) 'Governing through the brain: Neuropolitics, neuroscience and subjectivity' The Cambridge Journal of Anthropology, 32(1), pp. 3–23. doi:https://doi.org/10.3167/ca.2014.320102.

Toga, A.W. *et al.* (2015) 'Big biomedical data as the key resource for discovery science', Journal of the American Medical Informatics Association, 22(6), pp. 1126–1131. doi:https://doi.org/10.1093/jamia/ocv077.

Wright, J. (2018a) 'Chapter 11—seeing patterns in neuroimaging data', in C. Ambrosio and W. MacLehose (eds) Progress in brain research. Elsevier Imagining the Brain: Episodes in the History of Brain Research, pp. 299–323. doi:https://doi.org/10.1016/bs.pbr.2018.10.025.

Wright, J. (2018b) 'The analysis of data and the evidential scope of neuroimaging results', The British Journal for the Philosophy of Science, 69(4), pp. 1179–1203. doi:https://doi.org/10.1093/bjps/axx012.

# Clinical Implications

**Abstract** Scientific realism is the philosophical conviction that the posits of science—quarks, electrons, forces—are real, not just ways of accounting for the world. If true, we might take from this the view that a full story of the world and everything in it is most possible through the pursuit of scientific experimentation and observation. The truths of science, on this view, are the truths of nature. Scientific truths are true despite what anyone may actually think of them. This poses a problem for one of the more salient objects in our world—human beings. From the perspective of each of us, what's most notable about the world is that it's a place that we are in at a time. We can't help but experience the world in terms of our subjectivity, in other words. Even the simplest cases of perspectival relativism attest to that. From a person's point of view, a full story of the world given in objective terms will describe all but one fact; the fact of that point of view itself (or the 'I' that describes).

**Keywords** Scientific realism • Psychiatry • Precision medicine • Neurobiological reductionism • Neurostatistical reductionism

It seems important to us that our nature as perspectival beings be factored into any story of the world that purports to be complete and true. As already mentioned, however, a scientific realist might well believe herself to be in possession of a means by which to discover truth no matter what anyone thinks of it. Thus, anything that purports to be a real part of the world, subjective perspective included, must be explicable scientifically. If not, its status as a real part of the world ought to be doubted. As David Hume says in his *Enquiry Concerning Human Understanding*, if we have no empirical information regarding some question, "Then, commit it to the flames, for it can contain nothing but sophistry and illusion."

Gaining a scientific perspective on the subjective dimensions of lived experience would be not only of inherent interest, but also of great clinical importance. Mental healthcare is expensive. By 2026 in the UK alone, the costs including treatment and lost earnings are expected to top £88.45bn (McCrone, 2008). Mental healthcare also produces sensitive public health issues, that touch upon matters of personal, social, cultural, and historical importance. This is in addition to the distress of those suffering with mental illnesses and their carers. Mental illness creates considerable social stigma and alienation. Hopes for science in yielding solid diagnostic and treatment strategies for psychiatry might reasonably be high. Given the output and insight from neuroscience gained through datafication, this too could bring advances in psychiatry. Looking to the brain itself as the substrate of disorders manifesting cognitively and behaviourally as psychiatric problems promises a consistent basis for diagnosis, and for treatment. This would offer psychiatry a scientific basis that can be hard to see throughout present and historical iterations of the discipline. Indeed, there is a well-established movement in psychiatry that promotes the concept of psychiatry as 'clinically applied neuroscience'.

Creating a novel psychiatric discipline of clinically applied neuroscience would circumscribe clinical practices of the future, with personal, social, and public health consequences. It would serve to define new means of assessing mental illness, with implications for how we will draw boundaries between health, normal difference, and pathology in human rationality and behaviour. This could be expected to arise from data in neuroscience to produce new concepts of mental illness, newly scientifically accurate diagnostic tools, and precision treatment strategies.

Bringing the dashboarding idea—from above—into psychiatry highlights important issues about human identity, psychological wellness, illness, and normal variation. In an experimental paradigm, researching transitions from sleep to waking, for example, there are clear beginning and end points: wakefulness and sleep. Could data promise more subtle insights, shedding light on nuances of mental states so as to account for differences among anxiety, depression, dissociation, mania, paranoia, or any number of other such examples? On the other hand, in processing data as applied to brain function and evaluation of cognition and behaviour, it may be too easy to forget the persons involved. Reducing mental illness to neurobiological categories, or biostatistical parameters, is a big step with important philosophical and ethical considerations attached.

## Psychiatry as 'Clinically Applied Neuroscience'

In 2005, Thomas Insel, then Director of the US National Institute of Mental Health, and Remi Quirion, Director of the Institute of Neurosciences, wrote that, "...clinical neuroscience must be integrated into the discipline of psychiatry..." and that in the future, "...psychiatrists and neurologists may be best considered 'clinical neuroscientists.'" (Insel & Quirion, 2005) With this, they were advocating a movement that sought to promote the concept of psychiatry as *clinically applied neuroscience*. Creating a novel psychiatric discipline of clinically applied neuroscience would circumscribe clinical practices of the future, with personal, social, and public health consequences. It would serve to define new means of assessing mental illness, with implications for how we will draw boundaries between health, normal difference, and pathology in human rationality and behaviour. This would arise from an interaction between neuroscience and psychiatry which is hoped by advocates to provide new concepts of mental illness, newly scientifically accurate diagnostic tools, and precision treatment strategies.

The thought is that using the tools of neuroscience, like fMRI and EEG, dysfunctional neural circuits can be identified that underlie mental disorders. Conventional definitions of mental illness have arisen heterogeneously over time, leading to vagueness and 'fuzzy boundaries' between different syndromes. Psychiatry, as compared with other medical sciences, lacks definitive diagnostic approaches, and treatment pathways. The evolution of various editions of *The Diagnostic and Statistical Manual of Mental Disorders* (DSM) and *International Classification of Diseases* (ICD) (See Regier et al., 2013) has in large part been an attempt to remedy this anomalous relationship between psychiatry and medicine in pursuing, "consistent clinical descriptions of syndromes", and "specificity, that is ability to distinguish different types of problems" (Kirmayer & Crafa, 2014).

Using DSM or ICD as a guide for structuring clinical interviews, psychiatrists can ensure clinical consistency. Nevertheless, psychiatric models based on medical approaches in general encounter challenges. Whereas genetics, for instance, has served to improve approaches to cancer care, the kinds of genetic-environment-context-behavioural relations experienced by any individual in day-to-day life make the discovery of genetic bases for psychiatric conditions vastly complex. While it is apparent from studies that look into genetic variation and individual responses to drugs that genetics and psychopathology can be linked, the mapping of those links remains difficult. They may also be structurally limited, as many of these studies on gene association with psychopharmacological response are predominantly carried out on participants of European descent. This creates significant gaps in the transferability or applicability of results concerning gene association obtained outside of European-descended individuals (Dalvie et al., 2016).

Brains are much closer to human action, disposition, and behaviour than genes. A promising way to overcome the genetic shortcomings suggested by this problem in translation beyond European populations might be to move to the brain. This could be good *prima facie* justification for recommending the pursuit of clinically applied neuroscience as a focal point for psychiatry. This justification might go: Disorders in human behaviour, mood, disposition and so on are more likely to have consistent neural correlates across diverse groups of people as the brain is essentially involved in everything any human being says, thinks, or does. So psychiatry that focusses on the neural correlates of human behaviours could appear to have access to what's really going on 'behind' that behaviour.

Cognitive science in general has sought ways to account for *what's really going on* when people behave in ways elicited inner experimental conditions. It was already suggested that, following the split brain experiments, it seems possible than one person might know more about another's behaviour than that person might know themselves. Those were very uncommon circumstances, despite the apparent lesson to be learned. But in in the 1980s, Amos Tversky and Daniel Kahneman seemed to show that in ordinary circumstances of decision-making people's inner models of the world failed to line up with what ought to be considered objective standards of reasoning (Tversky & Kahneman, 1983). Through some decision task experiments, they were able to demonstrate that experimental participants would very often make irrational decisions. For example, one task involved reading the description of a fictional person called 'Linda'. Having read the description, participants were asked to choose from a list of attributes those they thought Linda was most likely to exhibit. The description included sentences like, "Linda is a bank teller" and "Linda is a bank teller, and a feminist". Many chose attributes that were demonstrably less likely than the others: they chose attributes that were conjunctions—'and' sentences—meaning they were suggesting combinations of two attributes were more likely than individual statements. Statistically, any conjunction will be a fraction of the possibility of either of its conjuncts.

The chances of rolling a dice and getting a five are 1/6. The chances of getting a four or a five are 2/6, or 1/3. Disjunctive probabilities, "or" sentences, are summed. This makes sense, as there are more outcomes being grouped into one roll. But the chances of rolling a dice twice and getting two fives means multiplying the probabilities. This makes sense because over the two rolls of the dice, fewer possibilities will count as what's aimed for, namely a five. So the probability of rolling a five AND a five in successive rolls is 1/36. Every one of Tversky and Kahneman's experimental participants who thought a sentence with more than one attribute listed for Linda was more likely than any sentence with just one attribute listed appears guilty of irrational behaviour.

This looks like another example of how an external observer can know better than an experimental subject about the correctness or not of how they might justify their reasoning in a specific set of circumstances. The participant who thinks that Linda is best described by a conjunction is objectively wrong, no matter what they think about it. In a sense, they aren't thinking properly about the case, despite how much conviction they may feel about their conclusion. Here, the link between 'right reasoning' and its assessment can be seen in analogy with the sort of assessments that could be made in psychiatry as clinically applied neuroscience. Beyond cognitive science research, in a clinical assessment, a psychiatrist would be tasked with observing a prospective patient's behaviour and avowals and deciding whether any anomalies amount to issues of concern. Apparent mismatches between perception or judgement of the world and objective factors could merit further investigation as potentially pathological.

Another interesting dimension to this line of research into human cognition was how experimental participants could be encouraged to make more objectively 'correct' decisions. This could be achieved through providing them with different descriptions of the tasks they were offered. Depending upon how the tasks were described, participant's rational decision making could be affected. For Gerd Gigerenzer (1991), the way minds appear to work is a result of how evolution provided human beings with their cognitive capacities. They track the likely evolutionary experience of human beings over time. Through understanding how human cognition evolved, the mistakes in the Linda case could be clarified and corrected for. Gigerenzer rephrased the tasks used by Tversky and Kahneman while repeating the general scheme, emphasising frequency rather than probability. So instead of saying, "Which is more probable?" and providing a list of attributes from which a participant should choose their answers, Gigerenzer provides the description of Linda and asks:

If 100 people fit the description of Linda, how many of those people are:

(i)  bank tellers
(ii)  bank tellers and feminists

Phrased this way, the cognitive errors seen in Tversky and Kahneman's experiments almost completely disappear. Gigerenzer's contention is that if a capacity is to be experimentally tested, the context of its origin ought to be included. Tversky and Kahneman's experiment abstracts too far from the ways in which human beings acquired a capacity for reasoning about conjunctions, and so their results appear to show irrationality.

What is important here is the way in which inner processes are considered to shape outer performances. The problem as seen by experimenters in this kind of context is to substantiate the divergence between objective propriety in reasoning and the subjective processes of reasoning going on 'in the heads' of participants. By noting divergences like this, further research can be done to establish ways to address them. This approach would appear to have a strong overlap with psychiatric practice in that each seeks to uncover underlying cognitive processes that relate to overt behaviour or action. In the cognitive science case, erroneous reasoning is apparently revealed and corrected through understanding the source of the error and adapting the task. In psychiatry, an underlying cognitive issue might be addressed through psychopharmacology or another such therapeutic response aimed at adapting behaviour or neurology. Tversky, Kahneman, Gigerenzer, and others' work all suggests that it is possible to get clear insights on gaps in human reasoning, and to correct for it through deeper understanding of how cognition has evolved and is implicated in everyday decision making.

We could easily see the role that neuroimaging could play in this more general, wider trend. If getting to the underlying, perhaps unconscious, motivation of behaviour can provide a means of understanding perceived shortcomings in that behaviour, then looking directly into the brain's activity might reasonably be assumed to be bedrock. This prompts some, such as Insel and Quirion, to look to the brain itself to ground a robustly diagnostic and therapeutic psychiatry. In focussing on the brain as a homogenous substrate of mental illness, it is hoped to develop a precision medicine approach, divested of complicated heterogeneity.

Since the early days of the twentieth century, following Karl Jasper's publication of *General Psychopathology* in 1913, psychiatry pursued several threads in approaching the understanding of mental illness. The aim was to create a 'gestalt' that captured the elements of the patient's position such that their condition could be understood, and a path toward treatment formulated:

"The Gestalt arising from a careful, detailed and thorough exploration is more than simply an aggregation of symptoms— just as the picture arising from a jigsaw puzzle is more than the sum of its pieces; it is a coherent picture of the patient's mind and the foundation of a valid clinical and diagnostic appraisal." (Schultze-Lutter et al., 2018, p. 448)

But this isn't so easily done, and gaps in the approach prompt concerns like those crystallised in Insel's position. Social and cultural values, personal and political histories and contexts are in play in the Jaspers-like approach. These factors are relevant to characterising mental illness in ways not seen in physical illness. Even in codifying specific symptom groups and terminology in DSM and ICD editions has limited application as editions change, making for clinical practices that are dynamic over time. But the brain is more stable in its form than these texts. It can be considered a consistent substrate, which the DSM and ICD editions could be seen as lacking. Psychiatric syndromes and definitions may change, and their boundaries change, but the brain and its functions persist as they are set in biological and physical terms. Its function can be studied, and over time a picture of how it ought to function can be developed. Rather than looking at overt performance of behaviour, we might look through this and peer into the brain so as to develop a picture of normal function. Where surface pathological performance is observed, it might be assumed that this reflects a deeper deviation from normal brain function. But the move toward a general picture of normal brain function requires the amalgamation of large amounts of information from a variety of contexts.

At least two interpretations of the task 'understanding the brain' are available. Firstly, one can seek an understanding of the brain as an object—a purely anatomical undertaking—exemplified perhaps in the drawings of Santiago Ramón y Cajal. Secondly, one can seek an understanding of the brain as somehow 'correlating' with cognitive faculties in general (Place, 1956; Smart, 1959; Gamez, 2014). This latter approach would be the means through which to achieve a model of the brain necessary for psychiatry as clinically applied neuroscience. If a neuroscientist of today wants to explore neural function, they might identify a phenomenon of interest, locate a brain region of interest, design an experiment to record changes in brain function while engaging that area of the brain in a phenomenon-relevant task, and assess the results they obtain. To make a model of the brain beyond each individual experiment carried out in each separate lab across the neuroscientific discipline, the data from everywhere would need to be amalgamated. Efforts to produce such a model can be seen to be underway already. The European Commission's Flagship research endeavour, the *Human Brain Project*, for example, combines supercomputing, analysis of human brain imaging, patient data, and lab work on human and animal brains with the intention of combining insights from all of these areas to understand or 'decode' the processes and functions of the brain. Essentially, from the scale of the molecule to (potentially) the socialised organism, an account of the brain that bridges these scales (Amunts et al., 2016) will be developed. This would be hoped to bring benefits for areas like diagnosis of brain disease, and psychiatric disorder just as clinically applied neuroscience ought to.

This necessity for creating types requires something that can classify huge amounts of heterogenous data according to sophisticated and maybe subtle parameters—maybe parameters not evident to those holding the relevant material. As already discussed, the role of techniques like CBDA play a role in making coherent accounts of a phenomenon from heterogenous data. The data available on brain activity and behaviour, distributed among research experiments and groups globally, and through time, look like a case study in machine learning, algorithmic sorting. They also appear to offer a sandpit for artificial intelligence techniques. Looking among all this processed data for a 'normal' would be a bedrock for psychiatry as clinically applied neuroscience.

From all of this data amalgamation, a model of a statistically 'normal' brain could be developed, and thereby institute a way of comparing deviations from 'normal function'. Philosophically speaking, this could be described as amounting to a 'type/token' distinction, such as described in Charles Peirce's *Prolegomena to An Apology for Pragmaticism* (1906, p. 506). The data-driven 'normal' brain is the 'type' of which any individual brain is a token. Token deviation from the type is expected, but in principle requires scrutiny as to whether is represents a pathology or merely difference. The 'type' is not real, but contains every relevant bit of token data, abstracted to construct a norm.

Whereas Jaspers' method approached the patient and sought a gestalt of subjective experience, clinical symptom, and potential course of treatment, here the 'gestalt' is a model built from swathes of data. This abstraction and proxification of relevant data can result in a dynamic and flexible model potentially including the specific dimensions of the target object. One recent tool has been built that seeks to act as "…an interactive open resource to benchmark brain morphology derived from any current or future sample of MRI data" and stand as a reference chart for brain morphology across a human lifespan (Bethlehem et al., 2022). Like a doctor might compare the height or weight of a person at stages in their lives to assess how 'normal' their development was, this brain morphology tool is intended to permit similar comparison for brain structures across a human life. This is ambitious, but in the context of this chapter we're imagining further and thinking of a reference for *normal function* with relevance for cognition and behaviour, not only morphology.

Let's imagine that a model exists that includes a range of 'normal' brain function, across the whole brain. Those working in clinically applied neuroscience—the psychiatrists of the future—have access to the model and surrounding expertise on how brain function relates in various ways to overt behaviour, disposition, and so on. Presented with a patient suffering from a potential psychiatric illness, the clinician of the future would set about recording brain signals relevant to the clinical situation and processing them in light of a range of received norms. Dysfunction on this account could be identified in terms of brain circuits not operating in expected ways, or not corresponding with activity across the whole brain in ways the model predicts. This would present a clinician with a solid, scientifically based, means of targeting 'misfiring' areas of the brain, and going on to see how these misfires correlate with behavioural, or cognitive issues. The place for brain signals in this diagnostic process is central, in providing hard data on neuro-functionality as a causal factor in cognition and overt behaviour.

Brain signals are much closer to thought and action than genetic information. Genetic traits serve to condition behaviour, for sure, but in a way that is much more diffuse than a motor neuron's action potential, indicating an imminent muscle contraction. Genetics might pre-dispose handedness, for example, but looking to the brain gives evidence for what a hand is about to do before that is observable yet as outward behaviour. This is a way of looking through or past outward bodily behaviour, or verbal self-reports of disposition or thought, in the ways anticipated in the discussion of the split brain experiments, and in Tversky and Kahneman's work. Here, though, the data is thought to report more clearly and decisively. With this kind of approach, the hope would be that a precision approach to psychiatry could emerge, rooted in neuroscientific insight.

## Precision Medicine

Precision medicine in general is an approach that seeks to leverage as much knowledge and information as possible about a patient and their illness in order to specify a personalised treatment plan for them. This might mean a personalised drug dosage, schedule for treatment, or specific kind of treatment. The precision medicine approach has had success in a variety of medical fields, like oncology and infectious disease, and so its pursuit in psychiatry can be considered desirable (Insel, 2014). There is a perceived limit in following symptoms in psychiatry, however, and a sense that this limit is a challenge specific to psychiatry. Symptoms of disease in cancer, for instance, can often be seen quite distinctly as attaching to a specific cause. A particular syndrome may attach to an identifiable kind of illness. With psychiatric illness, clusters of symptoms can appear as manifestations of physical illness, side-effects of drug treatments, or one of several psychiatric disorders. Insel illustrates this problem with respect to 'anxiety,' which from a clinical point of view might be a symptom manifested by a series of different causes. The hope is that with mental disorders identified as *brain disorders*, and a raft of neuroscientific approaches at hand, psychiatric disease can be targeted and treated very precisely. The vagueness and 'fuzzy boundaries' among the syndromes of DSM and ICD are replaced with scientifically-grounded definitions of brain-based pathology as mental illnesses, in a context termed the 'Research Domain Criteria' (RDoC). Given the unique challenges of psychiatry within medical science more generally, this appears promising. But psychiatry as clinically applied neuroscience is not without its own complexity.

Categorisations of disease from DSM or ICD are based in a process like discussion, formalised and constrained by professional experience and clinical knowledge built up over time. This has some well observed problems concerning consistency and validity of classification. Where these approaches appear to be most successful is in terms of producing a common vocabulary among a variety of psychiatric practitioners. In this vocabulary, the adequacy or inadequacy of classifications and symptoms, descriptions of syndromes and behaviours, can be stated and debated. This puts psychiatry in a discursive mode that is quite unlike the case of other medical contexts. The precision approach here described places the categorisation of symptoms, syndromes, and illness a step back, in the activity of the brain itself. This is more like other branches of medicine in positing a biomarker for disease. Pathology of brain activity points to disease, in this account.

According to Insel and Quirion, and in the general context of computational neuroscience, biomarkers for neurodegenerative and other diseases will serve to revolutionise targeting of illness in specific individuals. The ability to pinpoint exact treatments for specific individuals will in turn allow for a turn away from medical practices based in judgement calls about individuals' circumstances based in generic diagnostic criteria, and toward scientifically grounded methods for determining health, natural variation, and disease. Treatments will be rendered more effective, on this basis, meaning patients will be better served by efficient diagnostic and treatment strategies.

As should already be clear, however, the 'pictures' of the brain used in this approach are not like the sketches of Ramón y Cajal, or like photographs, but are complex statistical representations of neural activity. They are derived from brain signal recordings, processed by complicated algorithms, and can be represented in more or less visually informative ways. For a precision medicine approach applied to psychiatry, this might seem to be a positive move away from the vagaries of the discursive toward a scientifically grounded psychiatry of neuroscience. Patient consultation of course would remain a starting point in this approach. Categorisation of reported symptoms is still a necessary part of the process. But diagnosis and treatment planning could be sequestered more behind a veil of data processing rather than discussion. Data processing has its limits, but here it would appear to be a more objective approach to the brain and its function than a completely verbal encounter with a patient.

Aside from, or beyond, the specifics of a clinical-patient encounter, the precision approach to psychiatry would also enable psychiatry more generally to proceed along the lines of other medical fields and closer to scientific progress in general. The findings of psychiatric encounters based in this data-centric, neuroscientific approach could be expected to aggregate over time. Investigation of the brain and of psychiatry might progress hand in hand, with understanding of mental correlates of neural activity emerging at the same time. Exploring pathology would lead to greater understanding of normal function too, leading to improved brain models related to conscious experience. But with growing complexity, and recalling the basis of this notional future for psychiatry in data processing, the role of processing cannot be overlooked.

At least one element of a precision psychiatry based in applied neuroscience could be characterised as a brand of 'data phenotyping' for neurological and psychiatric disease. On the one hand, data processing of brain signals by algorithms, to extract interesting signal features, is a necessary part of this approach. This is the business of neuroscience as it happens now: specific brain signal features are examined as they relate to specific areas of investigation, like vision, memory, or emotion. A large part of this is 'signal sorting' in order to isolate relevant signals from more general brain activity. Norms for brain activity can be derived from growing datasets detailing signals in relation to experimental tasks and thereby comparison made among derived signals in order to assess their characteristics. But as already noted for psychiatry, there is the potential for a lot of complexity once evaluation of behaviour and pathology are brought into the picture. A much more complex and responsive kind of data processing is required, beyond signal sorting or comparison among instances of derived signals, relative to specific tasks.

If we fast forward to a future in which psychiatry as clinically applied neuroscience has already matured, we might expect psychiatric diagnosis to proceed according to a 'screen and intervene' model (Rose, 2010). In this instance that might mean that we might come to expect a set of neuroimaging investigations to be administered such that pathology would be detected and a course of corrective intervention recommended. The role of the psychiatrist in this paradigm would likely still include patient consultation. But if the ambition is to identify biomarkers for psychiatric disease as based in neurofunction, then a mature psychiatry-as-applied-neuroscience ought to come to resemble this more objective diagnostic situation. This kind of situation would require that brain function could be assessed as pathological, deviant within normal parameters, or sufficiently normal to warrant no further action. That would require signal sorting, neurofunctional comparison within the brain at hand, and according to norms across datasets. And all this *in general*, that is, not specific to a particular experimental paradigm. The brain's function, after all, is what's at stake in this kind of approach. Not the activity of a brain specific to, say, the investigation of specific brain function in terms of a sense modality, or memory, or emotional valence. This kind of complexity would require a step change in approach to brain data, well beyond the application of algorithms chosen for specific purposes. The processing here would need to be dynamic, responsive, and self-guiding. Brain data processing for the ultimate ambitions of future precision psychiatry would require the input of artificial intelligence.

The role of artificial intelligence (AI) in neuroscience and psychiatry is not yet dominant but is growing. It would be able to play a greater role but for a relative lack of data on which to train. AI requires huge datasets from which it can generalise and thereby 'learn' to classify novel inputs in ways that appear intelligent. It is exactly this sort of intelligent classification that would be required for a future precision psychiatry to become operational. This kind of AI processing of brain data would be the basis on which signal sorting and so on could be surpassed, and from a swathe of complex, dynamic brain data predictions could be made that could be connected with behaviour. Ultimately, AI processing of brain data would appear to be a crucial element in the future of precision psychiatry. With increasing interest in collecting brain data, not least since the 'decade of the brain' was heralded in the US and the EU began the *Human Brain Project*, the groundwork is being set for the kinds of huge datasets needed to train AI to work on neural data. Besides this boost for data extraction from the activities of living brains, and the potential this brings for brain modelling and for AI analysis of that activity, other dimensions of clinical practice too are open to datafication.

Clinicians have always collected data on patients, through observations, case notes, and more general medical research findings. All of this combines in specific clinical encounters to justify specific diagnoses in specific cases. If we recall "ChestX-ray14" at this point—that was the successful radiology AI that diagnosed hernias and lung shadows in x-rays—we might get a glimpse of what future precision psychiatry-as-applied-neuroscience would be like. Remember that CheXNet included clinical notes as well as imaging data, and so it was in some sense reproducing clinical decision-making as well as image analysis when detecting pulmonary disease. The AI system can reproduce the results of the medical reasoning, if not with the understanding that brings it about. That notwithstanding, the results of the system appear quite impressive. As such, a good role for such a system is in 'decision support' meaning clinicians could refer to the system's recommendations as part of their wider clinical reasoning on a case-by-case basis. Applied to clinically applied neuroscience, or the psychiatry of the future, one could see this as an ambition too.

The use of complex processing of neuro-data could help to inform clinical diagnoses and treatment strategies. This could not only remove some burdens from psychiatrists' workloads, but also possibly remove biases and correct for gaps in clinical knowledge. If every psychiatrist had access to a system based in data curated from worldwide datasets, one could imagine a more consistent and expert—possibly more accurate—psychiatric practice emerging as a result. In short, the use of an AI-based data-analytic psychiatry could make for a more precise clinical practice of psychiatry. In principle, any given psychiatric case could be treated in terms of every psychiatric case, mediated in collections of data smartly processed by complex layers of machine learning and inference.

The skill of the psychiatrist would still be a necessary part of this kind of advanced, data-driven psychiatric system. But it might be a skillset emphasising dimensions other than those seen currently. Aptitude with data science, for instance, would come more to the fore than qualities like those associated with a sympathetic bedside manner, or related interpersonal skills. In terms of precision medicine, however, this could be cast as a positive step. The vagaries of sympathy and interpersonal skills are hugely positive when applied well but are open to subtle biases. Clinicians, just like everybody else, may be prone to unreflectively value or expect specific kinds of reactions from people they recognise in some way or another. This can lead to specific forms of experience being privileged, and others missed. This is well known in terms of race and gender bias in medicine, including psychiatry. Through data, the hope is that what's human shines through without the distorting lenses of social judgement. But given what has already been said about the limits of data, this may not be as straightforward as it seems. Nevertheless, among many, the hope remains.

Another more radical hope for the future of psychiatry could include the idea that AI and brain data can be used in a more directly instrumental way. The data-centric approach surpasses limitations present in case study or cohort analyses by aggregating wide ranges of quantitative data from instances of imaging, for example. By analogy with CheXNet, qualitative observations too can be datafied leading to diagnoses of illness comparable with clinicians (though recall that example reproduced clinicians' weaknesses as well as strengths). A data-centric approach in which research and diagnosis are codifiable seems in principle to permit a more complete computational approach to medicine. The 'screen' of 'screen and intervene' would be complete in these terms. In the extreme, this would diminish the role of medical theories, and expertise, replaced by exploratory data science. In the specific case of psychiatry, moreover, there could also be scope for the inclusion of the 'intervene' part within the data processing.

Neurotechnologies include devices that can modify brain activity to achieve specific aims. In non-invasive devices worn simply on the head, this can be done through electrical stimulation or the use of magnetic fields, for example. Typically, such devices are user controlled. A user can specify what kind of neuromodulation that would like, according to the parameters of the device. It's then up to the user too to decide whether the device acts as promised. But there are also 'closed-loop' devices which can be set to produce a target state which the device itself then maintains. For example, for cases of severe epilepsy devices are being developed that would automatically monitor and intervene to arrest onset of seizures. The devices would monitor brain activity, and when overstimulation likely to overload the brain and produce a fit was detected, intervene by modulating brain activity through electrical stimulation. The stimulation would interrupt overloading and stop a fit before it happened. In principle, this would be a closed loop system as it would serve to screen and intervene all on its own, according to preloaded parameters about neuro-electrical dynamics and thresholds for problem instances. If we imagine these closed-loop systems in a psychiatric context, this could be a paradigm shift in the field.

It might be questioned that a device could really, on its own, be used in a psychiatric context. The role of dialogue and understanding, con-structing the gestalt discussed by Jaspers, would be completely absent. But given a background like RDoC, and the possibility in principle of a precision medicine approach for psychiatry, a closed-loop psychiatric neurotechnology would be a valuable prize. While there are doubtless complex cases in which psychiatric diagnosis and treatment are extremely fraught with disputed details of communication, mutual understanding, perhaps delusion and obsession, there may be others more amenable to neuroreduction. Some types of depression might be such cases, where neurofunction can be seen as leading fairly directly to behavioural, cogni-tive, or perceptual issues. Such examples might be cases for trialling closed-loop neurotechnological *theranostic* or *electroceutical* devices.

'Theranostic' devices combine therapeutic intervention and diagnosis, while 'electroceuticals' are treatment interventions based in electrical stimulation rather than pharmacology. Each of these technologies in principle exploit knowledge of the brain's electrical circuits in order to treat conditions or illnesses through administered electrical impulses. Devices of this kind present the potential for a therapeutic paradigm without a clinician in the loop. Their potential for precision medicine comes from their role in constantly calibrating neuroelectrical activity toward a given optimal state, rather than periodically administering drugs or therapy. These latter more familiar approaches are like those consid-ered vague and insufficiently objective by advocates of psychiatry as clini-cally applied neuroscience. The potential for theranostics or electroceuticals appearing in psychiatry is not yet fully developed, but it seems neverthe-less an apparent endpoint for the trajectory begun in promoting RDoC. Given growing understanding of how the brain's electrophysiol-ogy predicts cognitive and behavioural activity from neuroscience, application of that knowledge in technologies comes to look like an obvious *next step* and one in line with medicine more generally—consider devices like the pacemaker or cochlear implant, and ongoing exploration into artificial hearts.

Neuroscience is becoming an important source of information for psychiatric practice. Mental disorders are being understood more and more as based in *brain disorders* and therapeutic interventions increasingly include neurotechnologies. At some point in neurotechnological development, psychiatrists may become less important in terms of day to day administration of calibrating brain activity for therapeutic ends. Across the spectrum of activity, from psychiatric decision-support right up to theranostic and electroceutical, the role of brain data becomes more central. This is so much the case that novel, artificially intelligent data processing techniques will be required to cope with the volumes and variety of that data. This combination of AI, neuroscience, and psychiatry nevertheless requires close philosophical and ethical scrutiny.

If we can use technology to revolutionise psychiatry, through understanding the brain in new ways permitted by AI and neurosciences, we should surely grasp every opportunity. However, if we aren't completely sure that we can use technology in this revolutionising way, we ought to proceed with caution. Each of these areas has uncertainties—internal presuppositions, cross-disciplinary methodological tensions, inconsistent aims—that don't automatically cohere. As long as AI, neuroscience, and psychiatric uncertainties persist, their convergence in neurotechnological applications will raise questions about how we ought to proceed (Rainey & Erden, 2020). These are ethical questions with practical stakes.

## Ethics Concerns: Neurobiological or Neurostatistical Reductionism

Unless the interaction between neuroscience and psychiatry is carefully arranged, we risk an *ad hoc* approach that could be inadequate, or even harmful, in overemphasising the neural dimensions of mental illness at the expense of broader personal, social, cultural, historical, and political dimensions. It will be worth examining a little more about what neuroscience brings to psychiatry in order to develop a picture of some of the challenges and ethical issues in its uses for psychiatry and precision medicine.

Brain research that interests itself in providing inputs to behavioural discussions (e.g., psychiatric treatments, or to broader discussions of the human condition), ought to be interested in contextualised human behaviours. The approach to the human brain itself (and to related behavioural diseases) must be considered optimal when it is multilevel. To approach it at one level only is not sufficient for a complete description. But it can be sufficient for a specific goal. For instance, to describe depression at the molecular level is not sufficient for describing the complex nature of the disease, but it can be sufficient for developing new pharmaceutical treatments (Changeux, 2017). The relations between diseases of the brain and psychiatric disorders such as autism, depression or anorexia—if these are indeed all reducible to psychiatric disease—seem unclear (Conrad & Barker, 2010). This is part of why research in this area is undertaken in the first place.

One overarching question might be posed like this: With psychiatry as clinically applied neuroscience, are we trying to get a picture of 'normal' brain function as a *telos* against which any brain ought to be judged as within acceptable functional variation or pathologically deviant? If so, data will provide this only when very carefully curated, monitored, and regulated—and not without some wide-ranging justificatory investigation. If not, what *is* the role of data in augmenting psychiatry, all things considered? Perhaps, using a different account of psychiatry such as one based in welfare, data could provide an account of difference rather than deviation from a norm. Datafied pictures of varying capacity among different individuals could provide boosts for self-understanding. But this might not necessarily, not in every case, translate into a diagnosis of disorder, nor a plan for 'treatment'. This overarching point should be borne in mind while the rest of the chapter analyses the notion of clinically applied neuroscience.

As with any science, neuroscience raises epistemological, methodological, and practical questions: epistemologically, neuroscientific knowledge is not clear-cut, because tools like fMRI and EEG relying upon choices among data curation techniques, and complicated statistical modelling (Poldrack, 2006; Vul et al., 2009). Methodologically, neuroscientists may consider their own lab-based work quite removed from human behaviour, where they are mainly concerned with close investigation of neural circuits in animal models, for example. Neuroscience in general might be thought of as being curiosity-driven, or driven by wide practical goals as laid down by research funding agencies (Baughman et al., 2006; Goering & Klein, 2020). Psychiatry may have something similar to neuroscience in terms of being curiosity-driven, but it will also have a public health agenda, sensitive to socio-political values that neuroscience may not routinely consider. This highlights the practical question of how to put the 'applied' in clinically applied neuroscience. That said, interdisciplinarity in neuroscience is very common, especially in research projects.

Neuroscientists often work alongside medics, psychologists, computer scientists, electronics engineers, and patients, not to mention philosophers. But how such a rich interdisciplinarity plays out from instance to instance is not necessarily a given (Lélé & Norgaard, 2005; Wodak et al., 2005). Sometimes, each respective academic or technical discipline's theories and models frame interactions, with each respective discipline mapping common ground relative to its own assumptions. In other instances, problems are taken as central, formulated in tandem among diverse approaches, and addressed collaboratively. Especially when the foundation of a novel field in psychiatry, with its own models of disease, diagnosis, treatment, based in neurophysiology is at stake, this question of interaction is complex.

To make the claim that a disorder is real and treatable is to tacitly suggest a view on what order is. In the case of psychiatric disorder, in the context of clinically applied neuroscience 'order' is provided by a normal set of brain functions derived from massive amounts of data. Rather than Jaspers idea of 'gestalt' based in a mosaic of understanding centring on behaviour, and information on motivation and intention, this is a *data gestalt* derived from many instances of brain recordings, processed and modelled. A wider aim in this area of building data gestalts includes trying to understand how the brain functions by generating very detailed simulations. These simulations in the broadest, most ambitious sense, aim to account for "behavior, cognition and intelligence" (Markram et al., 2011). The idea is that from the simulation—which can be controlled, run many times, observed in detail at different levels—researchers might then extrapolate from the data-driven simulations to experimentally observed anatomy and physiology of neuronal activity. Brain research in this sense is used to light the touch paper on massive scale simulations, which then feed back into basic research again, forming a virtuous circle. Neuroscience feeds simulation via the data it produces, then simulation accelerates neuroscience in turn. Normal functioning of the brain can then be theorised and observed in simulation and *in vivo*.

But what 'normal' does this new paradigm harken toward? Standard approaches like seeking a 'gestalt' or as outlined in DSM5 rely on the idea of 'mental disorder' as a disturbance in cognition, over emotional control, or behaviour. Disorder in the case of a data gestalt comes from a mismatch between a simulation and in vivo behaviour, perhaps, or in a prediction derived from a simulation that is not realised in a specific brain under observation. Or, again, perhaps the normal brain function derived from the processing of data gleaned from many instances of brain recordings is a very complex average with acceptable limits to deviation. Disorder in that case would simply be deviation from the complex average in some respect that falls outside of acceptable deviation.

If the data collected is considered unvarnished—it is after all from non-voluntarily manipulable neuronal sources—it may appear to ground an objective natural shared and basic collectivity among human brain function, based in datafication of many neural recordings. What it misses is that any token deviation is a statement of the statistical provenance of the token norm itself. Its generation is a statistical operation upon the data collected from many tokens, across various experimental scenarios, times, and contexts used to generate the type as an aggregate.

This parallels the ubiquitous data collection for social science, including in its assumed neutrality. In this emergent 'we' from massive data, however, proxification of relevant features is necessary. Whereas sociology can resort to interview, fieldwork, and so on, to contextualise what is to be proxied, a parallel empirical resort in neuroscience is not clearly forthcoming. Neuronal network development can be affected by environment, as laid out in the epigenetic theory of neuronal development (Changeux et al., 1973). 'The environment', moreover, can include social and cultural context (Kirmayer & Gold, 2011). Overall, it can therefore be stated that 'brain data' is not a consistent set within the individual across a period of time. But while the context may explain the expression of difference in neuronal development, one could not infer the context from altered neuronal development. There may be no feasible way to tell what the origin might be of any specific deviation in one brain—normal development, pathology, social effects, cultural context. Averaged across all collections of brain recordings from which brain data are derived, a normal may be constructed that has the value that it does owing to obscure and diverse causes.

All of this would be based in a very linear approach from identifying norms of brain function, associating them with norms of cognition and behaviour, and rectifying observed deviations from these norms through intervention when the deviations represent pathology. But the inherent complexity of identifying pathology remains despite the move into data. Data as it appears in a general picture motivated by RDoC, of clinically applied neuroscience, may have a more scientific basis in some respects than the interpersonal fuzziness of psychiatry as we know it. But it retains the issue of having to identify 'normal' in a moral sense. It has to identify not just statistically normal patterns in data that are or can be associated with normal and normally variant behaviours. It must also speak to what is pathological, and thereby make judgement calls about whether it is *good* or *bad* to intervene on one set of brain activity or another. In setting a normal, moreover, and intervening so as to promote normal function in a specific brain, for a specific person, the data-driven perspective is still tacitly claiming that bringing that brain of that person closer to normal function is an improvement. The patient ought to be *better off* after being treated.

Where simulation of the brain is cited as a method as well as an output from research, moreover, data appears in the methods and output. Heterogenous data, processed and amalgamated, grounds the construction of a type from which norms are derived. When investigating specific brain activity, the token data generated from specific recordings is then confronted with the constructed, type-derived norm. In assessing the token activity in light of the type norm in order to make a judgement as to whether it ought to be considered within normal range, pathological deviation, or something else, the token is assessed in terms of the type. But there is no principled difference between one and the other. The token seen in light of the type could just as easily have been data assimilated in the construction of the type. To draw conclusions about 'the brain' from this complex of data would be to have data reporting upon itself by introducing a pseudo-deductive loop into an inductive context (Rainey, 2022). It isn't just the case that we aggregate more and more data in order to have random fluctuations cancel themselves out, because in this case data selection, processing, and aggregation are related to one another (see Fig. 1 in Chap. 2).

The datafication of brains through data-driven, clinically applied neuroscience runs the risk that human subjects are themselves to some extent *virtualised*. In being re-cast in data terms, it is not clear that human beings, with their various potential neuronal patterns, lesions, and psychiatric orders and disorders, remain the targets of psychiatry as clinically applied neuroscience. Even if it is the case that neuroscience's target is a specific neuronal pattern rather than the human being as such, the problem remains that virtualization can potentially affect even these particular targets through the manner in which data is acquired and processed. From this epistemological concern over the knowledge-contribution provided by the data, comes the ethical issue that human beings and their agency, the neuronal patterns in their brains, may be deformed by accounting for them in a way that is insensitive to their biological, social, and personal plurality. Relying on the type, derived from the mass of tokens, as a norm might be instructive in various ways, for various research aims. But it will not serve as a telos, or an end toward which any deviation ought to be steered. This is especially important where neuroscience would hope to inform psychiatry.

Whereas the standard DSM is based upon symptoms and behaviours, neuroscience-driven diagnostics could instead be based on brain data. It is among patterns in brain data that neuroscientific success might be judged. Where this data is a proxy, the lesions, defects, or disorders it may indicate could well be those of a virtualised subject, remote from any particular human being. The dangers here are manifest where these data are not further interrogated and contextualised such that they are seen as simply another tool in the diagnostic approach. There is a difference to be made in using data to investigate the brain versus investigating people and their behavioural or psychiatric disorders.

By contrast, a welfarist approach emphasises a relational approach to wellbeing, with psychological traits playing a more embedded role with respect to a person, their environment, and the relative advantages and disadvantages netted from a multidimensional interplay among elements (Roache & Savulescu, 2018). The welfarist approach, among others, critiques a tendency toward a biological approach emerging in DSM 5, and replaces it with an interpersonal and contextual appreciation of thought and behaviour. The biological tendency itself emerges as a reaction to a conception of psychiatry as too arbitrary, non-medical, or unscientific. With psychiatry as clinically applied neuroscience, a novel take on the same kind of issue is seen. Here, the appeal is not to medicine, biology, or relation and context, but to the brain as discernible through data. Data might squeeze out alternative points of view, like that of the welfarist or another, and will risk positing a sterile, statistical 'normal' based in an array of data but not necessarily one correctly representative of 'normal' in a more thickly evaluative sense—even if there can be such a thing. Normality will itself remain a contestable concept. Data won't help 'see through' to some obviously desirable substantive norm.

The therapeutic potential of neurotechnology is tantalising in promising individual, brain-based means of overcoming problems in mobility, mood disorders, cognitive impairment. It may also offer enhancement prospects. Nevertheless, it contains puzzling elements and open questions that make those promises seem highly elusive. The idea of 'mind'-controlled devices more generally, outside of a clinical context, is more widely seen as an interesting mode of engagement with technologies for consumer applications or creative pursuits. In a consumer context, away from clinical purposes, these technologies are exemplified by software or devices operated not with physical interaction or voice commands, but through direct brain control. Brain signals can be recorded by way of EEG headgear and processed as input to control a variety of applications. Brain data are the information that results from processing the signals that can be recorded from the brain. Brain controlled devices will produce, store, and reuse brain data for recreational uses. Issues arising in this non-research and non-clinical domain may appear less urgent than those relating to understanding the brain or diagnosing and treating its ailments. Nevertheless, the role of brain data beyond the lab and clinic should not be overlooked.

# Bibliography

Amunts, K. *et al.* (2016) 'The human brain project: Creating a European Research Infrastructure to Decode the Human Brain', Neuron, 92(3), pp. 574–581. doi:https://doi.org/10.1016/j.neuron.2016.10.046.

Baughman, R.W. *et al.* (2006) 'The National Institutes of Health Blueprint for Neuroscience Research', Journal of Neuroscience, 26(41), pp. 10329–10331. doi:https://doi.org/10.1523/JNEUROSCI.3979-06.2006.

Bethlehem, R.A.I. *et al.* (2022) 'Brain charts for the human lifespan', Nature, 604(7906), pp. 525–533. doi:https://doi.org/10.1038/s41586-022-04554-y.

Changeux, J.-P. (2017) 'Climbing brain levels of organisation from genes to consciousness', Trends in Cognitive Sciences, 21(3), pp. 168–181. doi:https://doi.org/10.1016/j.tics.2017.01.004.

Changeux, J.P., Courrège, P. and Danchin, A. (1973) 'A theory of the epigenesis of neuronal networks by selective stabilization of synapses', Proceedings of the National Academy of Sciences of the United States of America, 70(10), pp. 2974–2978.

Conrad, P. and Barker, K.K. (2010) 'The social construction of illness: Key insights and policy implications', Journal of Health and Social Behavior, 51(1_suppl), pp. S67–S79. doi:https://doi.org/10.1177/0022146510383495.

Dalvie, S. *et al.* (2016) 'Toward a global roadmap for precision medicine in psychiatry: Challenges and opportunities' OMICS: A Journal of Integrative Biology, 20(10), pp. 557–564. doi:https://doi.org/10.1089/omi.2016.0110.

Gamez, D. (2014) 'Are information or data patterns correlated with consciousness?', Topoi, pp. 1–15.

Gigerenzer, G. (1991) 'How to make cognitive illusions disappear: Beyond "Heuristics and Biases"', European Review of Social Psychology, 2(1), pp. 83–115. doi:https://doi.org/10.1080/14792779143000033.

Goering, S. and Klein, E. (2020) 'Fostering neuroethics integration with neuroscience in the BRAIN initiative: Comments on the NIH neuroethics roadmap', AJOB Neuroscience, 11(3), pp. 184–188. doi:https://doi.org/10.1080/21507740.2020.1778120.

Insel, T.R. (2014) 'The NIMH research domain criteria (RDoC) project: Precision medicine for psychiatry', American Journal of Psychiatry, 171(4), pp. 395–397. doi:https://doi.org/10.1176/appi.ajp.2014.14020138.

Insel, T.R. and Quirion, R. (2005) 'Psychiatry as a clinical neuroscience discipline', JAMA, 294(17), pp. 2221–2224. doi:https://doi.org/10.1001/jama.294.17.2221.

Kirmayer, L.J. and Crafa, D. (2014) 'What kind of science for psychiatry?', Frontiers in Human Neuroscience, 8. doi:https://doi.org/10.3389/fnhum.2014.00435.

Kirmayer, L.J. and Gold, I. (2011) 'Re-socializing psychiatry: Critical neuroscience and the limits of reductionism', in S. Choudhury and J. Slaby (eds) Critical neuroscience. Oxford, UK: Wiley-Blackwell, pp. 305–330. doi:https://doi.org/10.1002/9781444343359.ch15.

Markram, H. et al. (2011) 'Introducing the human brain project', Procedia Computer Science, 7, pp. 39–42. doi:https://doi.org/10.1016/j.procs.2011.12.015.

McCrone, P.R. (ed.) (2008) Paying the price: The cost of mental health care in England to 2026. London: King's Fund.

Place, U.T. (1956) 'Is consciousness a brain process?', British Journal of Psychology, 47(1), pp. 44–50.

Poldrack, R.A. (2006) 'Can cognitive processes be inferred from neuroimaging data?', Trends in Cognitive Sciences, 10(2), pp. 59–63.

Rainey, S. (2022) 'Datafied brains and digital twins: Lessons from industry, caution for psychiatry', Philosophy, Psychiatry, & Psychology, 29(1), 29–42. doi:https://doi.org/10.1353/ppp.2022.0005.

Rainey, S. and Erden, Y.J. (2020) 'Correcting the brain? The convergence of neuroscience, neurotechnology, psychiatry, and artificial intelligence', Science and Engineering Ethics, 26(5), pp. 2439–2454. doi:https://doi.org/10.1007/s11948-020-00240-2.

Regier, D.A., Kuhl, E.A. and Kupfer, D.J. (2013) 'The DSM-5: Classification and criteria changes', World Psychiatry, 12(2), pp. 92–98. doi:https://doi.org/10.1002/wps.20050.

Roache, R. and Savulescu, J. (2018) 'Psychological disadvantage and a welfarist approach to psychiatry', Philosophy, Psychiatry, & Psychology, 25(4), pp. 245–259. doi:https://doi.org/10.1353/ppp.2018.0035.

Rose, N. (2010) '"Screen and intervene": Governing risky brains', History of the Human Sciences, 23(1), pp. 79–105. doi:https://doi.org/10.1177/0952695109352415.

Schultze-Lutter, F., Schmidt, S.J. and Theodoridou, A. (2018) 'Psychopathology—a precision tool in need of re-sharpening', Frontiers in Psychiatry, 9, p. 446. doi:https://doi.org/10.3389/fpsyt.2018.00446.

Smart, J.J. (1959) 'Sensations and brain processes', The Philosophical Review, pp. 141–156.

Tversky, A. and Kahneman, D. (1983) 'Extensional versus intuitive reasoning: The conjunction fallacy in probability judgment', Psychological Review, 90(4), p. 23.

Vul, E. *et al.* (2009) 'Puzzlingly high correlations in fMRI studies of emotion, personality, and social cognition', Perspectives on Psychological Science, 4(3), pp. 274–290. doi:https://doi.org/10.1111/j.1745-6924.2009.01125.x.

# Wider Markets

**Abstract** The global market for neurotechnology products had been estimated by various market prediction bodies at around $3 billion in 2020
and has been predicted by some to grow to around $15 billion by 2024
and $19 billion by 2026. Neurotechnology includes medical and research
applications, but also consumer devices. Right now, technologies are
already on sale that claim to have 'wellness' or health benefits, through
user recording, monitoring, and manipulating of brain activity. In consumer cases, a target brain state may be identified as being one promoting
concentration, or calmness, for example. The general electrical activity of
the brain can be recorded and intervened upon by the system in order to
promote that target state. There are issues about how this can be done
safely outside the lab setting, without supervision, and what long-term
effects there may be (Coates McCall et al., Neuron 102(4):728–731,
2019). There are also concerns about how these systems could be open to
hacking. Remote control of another's neuromodulation device could permit unprecedented degrees of manipulation (Pycroft et al., World
Neurosurgery 92:454–462, 2016). Established views on the role of data,
personal, and public accountability will likely require updating in light of
such developments.

© The Author(s), under exclusive license to Springer Nature Switzerland AG 2023     **93**
S. Rainey, *Philosophical Perspectives on Brain Data*,
https://doi.org/10.1007/978-3-031-27170-0_4

**Keywords** Consumer markets • Wellness • Consumer neurotechnology • Big tech • Markets in data

Brain data beyond the clinical context will emerge in 'wellness' applications, while consumers use 'mind' operated technologies. Such technologies include brain-to-text typing apps, and EEG-controlled drones. For those with access to it brain data will be a valuable asset—but access ought not too simply to be equated with ownership. Already, a variety of research, consumer, and news media reports have indicated shady practices by companies like Google and Amazon in using voice assistant data to track and profile users. Customers are able to block such tracking only at the expense of disabling their device's functionality. With it enabled, however, data is derived from input like voice commands in order to predict behaviour. These predictions are then used to raise advertising revenue based on sales of advertisements bought and sold in real-time, aimed at specific users.

Over time, detailed profiles can also be constructed. While these functions are known about, they are probably not considered 'the function' of things like voice-activated assistants. As such, the data being derived from customers' use of the devices is auxiliary to the main presumed function. It is accessible to Google, Amazon, and others, but it is difficult to reconcile it as *theirs* when the general use of voice assistant technology is considered. For many users, their voice assistant is there to play music and make shopping lists. For the technology companies behind them, this is merely a front-end for generating advertising revenue and consumer profiles. The fidelity of these profiles with respect to ongoing consumer behaviour is part of their value. Brain data would, for these companies, represent premium data for profiling as it relates directly to the person, unmediated by speech or other behaviour.

Facebook's (now 'Meta') intended brain-computer interface would not only record and process brain signals, but also associate the data derived from them with detailed social media activity. This would allow neuro-profiling, with brain data appearing to offer unvarnished insights into a person's behaviour and cognition. Such data may be very intimate, more akin to a diary entry than a statement for public consumption. The inferences drawn from it—whether they end up being accurate or not—could include very sensitive information. These might be akin to having a diary entry not only read, but analysed for patterns thought to be indicative of neuroses, hidden dispositions, or other private or unconscious dimensions of the self. They could include guesses about mental health, or concerning susceptibility to 'emotional contagion,' as already explored scandalously by Facebook (Kramer et al., 2014). According to a report by Privacy International, mental health data is already being bought and sold (Privacy International, 2019).

Targeting information at specific individuals or groups based on neural data and the inferences drawn from it, including those that could concern perceived vulnerability, would represent a new front in data-driven marketing or political campaigning. This would enable novel, more sinister, and perhaps harder to detect, forms of manipulation. Market forces cannot be relied upon to regulate this activity as those holding the data-assets in the market would stand to benefit from leveraging data-centred practices, while consumers would have limited options to choose for or against those practices. We cannot afford to have a large-scale brain data industry emerge without intense ethical and regulatory scrutiny. How this data-ecosphere ought to evolve—if it ought to at all—should be a matter of concern for all of us, given the high stakes.

## Consumer Markets in Brain-Based Technologies

An era of brain-controlled devices is emerging. The global market for direct to consumer (DTC) neurotechnology products is set to grow. Right now, technologies are already on sale, both on the largely unregulated consumer market as well as on the medical device market-that claim to have 'wellness' or health benefits, through user recording, monitoring, and manipulating of brain activity. Beyond the lab and the clinic, DTC devices represent an emerging front in brain recording. Consumer neurotechnologies may be recreational, like drones flown 'with the mind' (Blain, 2019). They may be associated with enhancement claims, to do with aiding cognitive ability, or part of the 'wellness' industry to promote meditative states, or aid in meditation (*Muse—Meditation Made Easy*, no date). All of this involves large-scale brain recording, with troves of data being amassed in the sorts of large databases apt for algorithmic processing.

Consumer neurotechnologies include devices that are designed to detect, record, process, and utilise brain signals. Owing to their target market, they are typically non-invasive, and don't necessitate neurosurgery for implantation. This would be a pretty immense barrier to product uptake—though plausibly not a total one as transhumanists may view a move into cyborg living as a desirable step. Indeed, Elon Musk's Neuralink is intended as a brain implant to be surgically placed in the brain, and it has a sizeable group of enthusiasts. Whether the enthusiasm persists as the device transitions from a largely speculative affair as of early 2023 into something more tangible remains to be seen. Maybe if the prospect of actual brain surgery for the device to be implanted becomes more realistic, interest will dissipate. In general, though, consumer neurotechnologies will be centred on ease of use, and adapted to fit with the body and standard patterns of day-to-day life. They will aspire to be relatively transparent to the user (to which we will return shortly, in discussing the potential ubiquity of consumer neurotechnology recording). All of this is aimed at securing a large consumer uptake of the technology, and maximisation of use.

The headsets developed by companies like Interaxon, Rythm, BrainCo, Kernal, and Neurable are each external devices, and appear to be designed with users in mind. Interaxon's *Muse* and Rythm's *Dreem* are each supposed to aid in sleep and so are intended to be unobtrusive enough not to inhibit rest. BrainCo's intention is to see its headsets in classrooms to aid in education, and such they ought not to be a distraction themselves. Kernal's device looks the least ergonomic in the sense that it is quite large. But the company's ambition to 'write' memories from the brain onto a chip through recording hippocampal activity probably mean it expects users to engage with the technology, and so it can afford to be more obvious. Neural aims to produce a headset that will act as a platform to allow control over and interfacing with other devices. Similarly, familiar Tech companies like Google, and Facebook intend to produce headset applications to allow interfacing with computers for things like hands-free typing. Again, these devices would seem to need to be inconspicuous so that they can become quite invisible to the user.

If you were to check the publicity around consumer neurotechnology, it would seem pretty clear just how these devices work: they are controlled with the mind. If we were to be stricter in evaluating things, this would be somewhat overstated. While it is true that, say, a neurotechnology-controlled drone would be piloted using something like Neurable's platform by thinking commands like 'up', 'down', etc., this is calibrated with neuroelectrical activity. The 'thought' correlates with electrical activity that subsequently is matched with the command. If you wanted, you could calibrate the device to raise the drone upward when you thought 'down' and it would calibrate that way. Or you could think 'orange' or 'antidisestablishmentarianism' or anything you liked. As long as the device can be calibrated to a distinct pattern of brain activity, it will function. So it's not 'the mind' that controls the device though one uses thoughts to produce effects—mental *content* doesn't matter. But what of devices like the Muse headset or Rythm's Dreem, that claim to modify brain activity? It would appear that they have some sort of effect on the mind. Again, though, with a closer look these devices operate based on a kind of 'neurofeedback' approach.

Brain activity is monitored and, where deep sleep is the goal, hyperarousal is targeted as a state to be inhibited. When such arousal is detected in the electrical activity of the brain, sounds might be played that are intended to promote calmer states thereby not allowing the arousal state to take over. In this case, the brain is monitored but not directly affected by the device in order to promote desirable brain activity patterns. Rather, auditory stimulation is deployed to do that job. More generally, neurofeedback might be thought of in the same way that a heart monitor could be used in a piece of gym equipment. One might decide that a target heart rate of 145 BPM is desirable. While using the gym equipment, the user can increase or decrease their effort, and increase or decrease the equipment's resistance in order to maintain a steady 145 BPM. In the case of neurofeedback, the target is neuroelectrical activity rather than heartbeat, but the principle is the same. Adapting one's activity so that a desired state is promoted allows some kind of control over brain activity. Or, in the case of Dreem, auditory stimulation is produced by the system to bring about a relaxed state.

BrainCo's device has a similar mechanism, based in neurofeedback. In this instance rather than monitoring one's own brain activity, or having a system produce auditory stimulus when specific patterns are detected, here a teacher plays that role. The idea is that while teaching a group wearing neuromonitoring devices, that group's neurofeedback can be followed by the teacher who can modify her approach depending upon the sorts of readings that she sees. As long as the students remain in states conducive to learning, things can proceed. But where distraction, or something else likely to inhibit learning begins to emerge, a shift in approach can save time that would be wasted in providing teaching for unreceptive students.

Auditory stimulation or changing teaching style based on observed brain activity patterns is clearly not quite the same as neurostimulation to change brain states. Neurostimulation would include electrical or magnetic fields being used to directly modify the behaviour of neurons firing in the brain, as considered when the discussion was about clinical examples of precision medicine neurotechnological devices of the future. Recalling the wider context of research applications for neurotechnology we might yet still wonder about just how a neurofeedback approach could work. Under research conditions, a huge amount of control is exercised over the parameters relevant to a specific instance of neural signal recording. If neural response to visual cues, for instance, are to be measured then it is important that signals produced in response to the stimulus are recorded. Not just any recording will do. This means timing is essential, in order to maximise the chances of associating neural responses with visual stimulus. Choices regarding where to record signals are also vital. Because electrodes in EEG headgear can record in very high temporal resolution, millisecond by millisecond, they are desirable in capturing fine-grained changes. But they lack a corresponding spatial resolution not least because each electrode placed on the scalp can record from many neural signal sources. Placing electrodes carefully to maximise the possibilities of recording relevant signals is important. In a lab, under research conditions, this task will be undertaken carefully. Outside of the lab, in day-to-day use of a consumer BCI, what can be said of these kinds of conditions?

In terms of timing, there is little sense to be made of recording specific responses to cues as was central in research neuroscience. In research, the aim is to correlate precise measurements of neural activity with controlled stimulus. In a case where neurofeedback is to be used to, say, promote a meditative state the user need only monitor their neurofeedback and try to react to it in such a way that they either feel more meditative or that their neurofeedback indicates a more meditative state. The correlations sought using this sort of application of neurotechnology are between subjective phenomenal states and neurofeedback, rather than any measurable response to stimulus. The nature of the neurofeedback itself is that of a processed signal already. The electrical signal recordings made by a user's headgear will have been sorted according to some criteria built into the software and algorithms of the specific—in this example meditation—application. Whatever has been programmed into the application's software as relevant to the task at hand will be presented to the user for their use. Moreover, if something like 'a more meditative state' is sought, this might mean different things to different users. Rather than needing to verify a consistent correlation between visual stimulus and neural activation, in an example case like this, already processed signal is assessed by a user according to their judgement about how meditative they have become.

Given software will already have processed recorded brain signals as they are presented to a user in a context of consumer BCI applications, it might be thought that electrode placement would be of special importance. After all, capturing electrical signal produced by the brain is not the difficult part. Sorting the complex whole into relevant constitutive signals is where the challenge lies, even in research and clinical settings. Algorithms for sorting signals have their work cut out, so optimising their input by optimising the source areas they record is a good idea. In a consumer headset, electrodes are unlikely to be as carefully placed as they are in a lab or clinical setting. Consumer headsets are ergonomic, as in they are designed to fit the head nicely. But each head is different. And each time the same person places their headset on their own head, it might be a little different to the last time. Given the high density of neurons per cubic millimetre of brain, and the differences in topography, thickness, and density of different parts of the skull, scalp, and cortex, a little difference in placement could have a large effect on recording source. And that's not to mention the dynamics of the brain itself as it moves around inside the skull, and syncs with breathing and cardiac rhythms.

The picture emerging here of neurofeedback recording outside of a lab or clinical setting is a complicated one. Signal sorting algorithms will be pre-selected and applied to signals as they are recorded from the user's headset. Not being task-dependent, the user will be confronted with the variety of stimuli normally associated with daily living. Attention cannot be assumed to be undivided either, with focus perhaps shifting among the different sensory inputs and mental contents. The recording electrodes cannot be assumed to be well placed over brain areas of relevance to some task, nor consistently placed from one session to the next. The signal sources, the neurons, will also be moving targets as neurotechnology users rove around. Satisfaction with the results of the devices used under these circumstances ultimately relies on the judgement of the user. None of which means that consumer BCIs couldn't 'work' to a satisfying degree. These devices, after all, can reap the benefits of ever more sophisticated basic neuroscience, supplying more and more clever algorithms and physical technologies.

In terms of signal sorting, for example, lessons can be drawn from astronomy and the use of algorithms there to distinguish signals from separate but very distant stars. In the astronomical case, signals from different very distant galactic entities arrive at terrestrial sensors superimposed on one another. Using complex but codifiable mathematical approaches, the characteristics of the tangled signal can be drawn out and the constituent signals derived. A radio telescope can then 'see' two distant stars that would otherwise appear to be a single complicated signal.

Learning from this, neuroscientists can create algorithms to disentangle complicated neural signal recordings to discern separate sources. Neurons aren't distant, like the stars, but their spatial density makes for a comparable problem in signal-sorting. From many lab-based experiments made under carefully controlled circumstances, algorithms can be developed to do this signal sorting in a way that allows for use in general circumstances. For a company designing a consumer BCI, this allows the possibility to produce a device that may not require the precise conditions of a lab-based application. By building on the advances of basic neuroscience, neurotechnology companies can afford to have looser technological approaches and nonetheless make reasonable claims about their products' ability to sort signals, thanks to powerful and well researched algorithms like those whose characteristics resemble an astronomical star-signal sorter.

With a powerful algorithm, electrode placement can come to mean less too. Where algorithms are well trained, they can likely work well enough for consumer purposes, such as neurofeedback applications whose verification principles are aligned with expressions of satisfaction. Ditto where good sleep is at stake: some cases of 'good sleep' will be achievable through the activities of a neurofeedback device designed to produce soothing auditory cues. Others might be achieved through a placebo effect induced by simply having the device. But a different angle to all of this consideration of consumer BCIs can be discerned from which these BCIs are seen as important signal gatherers. As stated already, recording brain signal is the easy part, so this is something consumer BCIs will be able to do. What should we think about this? Having seen the elegant, ergonomic, attractive designs of consumer BCIs headgear it might be obvious that masses of signal recordings aren't being stored locally. While it's true that the costs of capacious digital storage has dropped a lot in the past decades, as has size, it is still the case that including storage on headgear introduces design challenges. Including hard drives of any kind could increase cost and weight, leading to higher prices and less comfort. Moreover, recording to a drive requires some kind of software and maybe additional chips are needed in order to handle extra processes for recording signals to local storage. More local processing also produces more heat, and draws more power, perhaps leading to a less comfortable device. It might even mean a less safe device if heat levels are high enough to cause discomfort or burns. This all suggests that an ideal set up for a consumer BCI device would include a cloud service.

Cloud storage for BCI data would mitigate the kinds of problems just mentioned. It would also allow for data amalgamation, and the processing of vast arrays of data from all device users. These would represent efficiencies from a neurotechnology company's perspective. Besides this, the store of aggregated data would be an asset. Especially owing to the ways in which algorithms can be improved through more and better training, neurotechnology companies with vast troves of brain data would have much of the material in place to develop or improve signal sorting algorithms. With ever-growing brain data troves, those companies could also secure a future for themselves as a source for developing and improving signal sorting algorithms.

If we imagine a company like Google or Facebook/Meta curating a brain data trove through their users' interaction with proprietary neurotechnology applications, the scale on which data could be collected would outstrip anything a research lab could replicate. It was also suggested earlier that one of the reasons AI plays less of a role in neuroscience and psychiatry now is that AI needs vast amounts of training data from which to build functionality. If these vast amounts of data are to come about soon, it is likely to be from the widespread adoption of consumer BCI devices connected to cloud storage facilities. Certainly, this is more likely to produce *big brain data* than, say, academic research studies. Neurotechnology devices as signal gatherers could have an important foundational role in opening research and clinical sectors to a deepening reliance on data-driven technology through becoming a sandbox for algorithms, and in providing a basis for training neuro-oriented AI.

Already, with the emergence of direct-to-consumer neurotechnologies, brain data are being created outside of any clinical or research context. Currently available consumer neurotechnology uses brain data to monitor neural states, control computer game avatars, or to fly drones. In each case, these endeavours will produce, store, and process troves of brain data. The devices themselves will likely find a market and satisfy many users. They will also likely open new markets in data connected to the ways in which, besides any specific applications, consumer neurotechnology will act as signal gatherers furnishing general databases of brain signal recordings. The subsequent uses to which these databases might be put will include developing algorithms for better consumer neurotechnology functionality. They might also serve to refine research signal sorting algorithms and develop new possibilities in AI and clinical applications like those sought by advocates of precision medicine in psychiatry as applied neuroscience.

# Consumers' Brain Data

A lot of interest and ambition attaches to the burgeoning consumer BCI sector. Whatever else consumer grade applications may be able to do—induce calm, enable better learning, control other devices—they are each of them brain signal recording devices. And given the very basic requirement to sort otherwise superimposed and tangled signals from one another, processing of brain signal recording is a must. In this way, any consumer BCI should also be regarded as a brain data generator. Ought this to be a concern, especially if it were true that some applications promised more than they delivered? Perhaps it isn't really a problem if brain recordings and data representations of neural activity are collected and stored via consumer BCIs if they basically amount to little more than useless collections of ones and zeros in a cloud server. Given what was seen in the discussions of techniques like compressive big data analytics, and the ambitions for precision medicine approaches in psychiatry as clinically applied neuroscience, it could be a mistake to be so blasé.

For those with access to it, brain data will be a valuable asset. This data will be the means of producing new and more powerful algorithms, by providing the training material required to boost automated processing's effectiveness. This will have impacts upon basic neuroscience by providing more choice among more deeply trained signal sorting functions. This is turn will enable advances in understanding the brain. More effective algorithms will allow bolder hypotheses to be advanced and tested in lab-based trials exploring how the brain works. With advances in basic neuroscience could come more ambitious approaches to drug development aimed at modulating brain function. This could have strong clinical appeal, and perhaps good ethical justification. The more detail that can be leveraged from experimental brain recordings, the more good might come from the increased understanding gained. As already suggested, this same line of progress could also be seen as opening the door for more involvement of AI in research and clinical applications, especially where neurotechnology is used. With a deeper understanding of brain function gained via powerful algorithmic signal sorting, and AI-enabled brain signal monitoring, a boost to theranostic device development could be expected.

This notional, possible future trajectory encompasses neuroscientific research, neurotechnology development, and revolutionary changes in clinical care. It is speculative and futuristic, but it is not unrealistic in terms of the ambitions many have concerning brain data. The market is growing, and people like Elon Musk use their platform to hype the prospects whenever they can. But the way the data here are discussed sounds somewhat generic, as if they are somehow a natural resource that have arisen by chance. In the context of consumer BCIs, brain data are realised through consumers' use of their devices, not by some force of nature. Given the paradigm shifts latent in the collection and exploitation of brain data, this is due some further consideration.

In the hands of private companies, with economic interests of their own, there will be great incentive to treat consumers' brain data as an asset to be exploited like any other. In the sense that brain data are essential to the functioning of consumer neurotechnology applications, it is akin to a physical part, like an electrode or circuitry. Brain data could be considered a part of the device. But it is nevertheless generated from the physiology of a user. In many other contexts, where data is derived from people directly, data protection laws come into effect. Europe's 2018 'General Data Protection Regulation' (GDPR) seeks to place control of one's own personal data in one's own hands. Under this regime, personal data can be used by a person to buy goods and services, but they must typically remain in control of that personal data.

If a company were to have your personal data on file, and you no longer wanted them to store it for some reason, you can request the deletion of the data. Similarly, if a company wishes to collect personal data for some reason, they must go through procedures to specify what data they want, why, in what quantities, and how they will secure, store, and erase it. At any point, a person whose personal data is involved may exercise the right to know about it and, as described, have it removed. On a scheme like this, it would sound like consumer neurotechnology users, not providers, would have greater claims to ownership over the data. This being so, users might have some kind of claim over the anticipated advances enabled by way of brain data aggregation and use. There appears to be a gap in the GDPR when it comes to brain data, however.

In the Regulation, 'personal data' is any data from which a person can be identified. Data from which no such identification is possible is regulated far more lightly. Anonymised data, or very well pseudonymised data, can be used more easily than personal data can in a variety of contexts. But despite experimental claims about the biometric potential of brain recordings, it is not clear if they really offer a means of identifying persons in the same way a fingerprint, name, or address could. It is very unlikely that those who formulated the GDPR, across 7 years of literature review, negotiation, and consultation, considered an advent of widespread consumer brain signal recording. But the potential gap regarding brain data in the GDPR may cause concerns in such a context.

If researchers are right that EEG recordings, for example, can be used to predict psychiatric disease by way of examining neural oscillations, or diagnose alcoholism by way of charting brain connectivity, they could be used to highlight anomalies in brain function very sensitive for those individuals (Buzsáki & Watson, 2012; Bae et al., 2017). This being so, not only could a person potentially stigmatised by a brain recording, but so too would a potentially sensitive dimension of their life be revealed. Typically, health data is considered deserving of special protection in data privacy contexts. Perhaps it is obvious why, but one part of this includes this possibility of stigmatisation based on a real or perceived illness. If brain data were able to be used to predict psychiatric disease or diagnose alcoholism in individuals, and potentially reveal information about their health status, this would be of concern.

Another issue stemming from this gap would be the right to have personal data removed from a system. The ways in which brain data could contribute to advances in neuroscience, neurotechnology development, psychiatry, and so on, derive from aggregation of data and re-processing. This re-processing could come from applying something like compressive big data analytics, or from using data to build models to train AI for specific tasks, like the way CheXNet was trained on a corpus of clinical notes and images. If you have allowed your brain data to be used in training a deep neural net (DNN), in what sense is 'your' data still 'in' that neural net? In the development of the system, your data along with many others' contributed to the probability weightings made among varieties of inputs, hidden layers, and outputs. Now, as the system operates, it has drawn upon the corpus of data in which yours resided. If you ask that your data be deleted, this would mean its removal from the corpus. But since the DNN is already up and running, and changing according to its operation, it isn't clear what 'removing your data' in this context would mean. Recall that with reading the brain to map semantic classification, the focus on data allowed the development of models going beyond specific results to ground generalisable approaches. In this context of a brain data market including consumer BCIs, the models not only exist, but are for the material gain of others. These others include big tech companies, their competitiveness, and their profit margins. BCI users could find themselves instrumental in creating profitable models through use of their brain data, but unable to meaningfully remove their data from that model should they wish. This might be another way in which a right to have personal data removed is frustrated in a context of brain data.

The nature of brain data ownership leads directly to concerns about gaps in regulation. Even if data ownership was clearer, there is still a question about how clearly the idea of trading in brain data can be formulated. In the imagined case of the DNN just mentioned, it would not be clear to anyone exactly what the outcome of developing such a system might be. The kinds of advances hoped for via a brain data revolution are open ended—or they might come to nothing. In what sense can a valuation of 'brain data' be made such that it makes sense to discuss trading in it? It seems more like a lottery ticket than a common or garden good, so how could a sensible price be formulated?

This question, in fact, points to another dimension of this challenge relating to power. If there is to be a brain data revolution such that advances are unlocked across the variety of neuro- fields, it will come through data aggregation and processing on a large scale. There are not many entities capable of providing the infrastructure and expertise to actually leverage brain data. These will be large technology companies, or well-funded private enterprises. In this case, price-setting comes as part of a buyer's market, speculating on a future of huge financial yields on small initial layouts. We might even see the emerging DTC BCI market in terms of this 'small initial layout.' Relatively inexpensive, widely available consumer devices of uncertain capabilities beyond brain signal capture and data production might look like the establishment of the brain data market already. These questions of ownership, data protection, privacy risk, and value may be passed over before they have been clearly posed.

Brain data collected, aggregated, processed, bought and sold on a basis of private and market interests could be the backdrop for some quite predictable future developments besides anticipated advances in research and clinical contexts. These are largely based in possibilities that arise from the sheer volume of data that could be amassed, the lack of scrutiny with which it might be faced given regulatory gaps, and its appearing in corporate rather than scientific contexts. Moreover, many people want to believe in the explanatory power of the brain. Whether it is true or not, it is widely believed that neural explanations have a special quality as conversation-stoppers: The idea that once you hit upon a neural explanation of something or other, you have hit bedrock.

Already, back in 1990, Barry L. Beyerstein wrote about the widespread susceptibility to 'brainscams' which sought to offer ultimate advice for wellbeing, gaining credibility by using neuroscientific jargon (Beyerstein, 1990). Philosophical questions about minds and brains have been around for many, many years. But when presented with stories about the brain as an explanation for some bit of thought or behaviour, a lot of people seem convinced pretty fast that the brain really does explain whatever is at stake. Fascination with the brain is understandable—it's so complex, mysterious, and central to everything we think, say, and do. It's a short step from this to a generalised faith in neuroreductionism and identification of brain claims as explanatory of mental or personal claims. Your brain *makes you* think, say, and do x when y. Strikingly, even those who are somewhat more expert aren't immune to this. McCabe and Castel asked their research participants to evaluate news articles on scientific topics, different articles but conveying the same news (McCabe & Castel, 2008). For many of the participants, articles that included images of brains were thought to be more credible than those that used bar graphs. Other researchers found that cognitive neuroscience students early in training too tended to rate explanations of psychological phenomena more highly when those explanations were accompanied by superfluous neuroscientific information (Weisberg et al., 2007). This suggests their paper's titular *seductive allure of neuroscientific explanations* is not straightforwardly neutralised by training and education.

A market in consumer brain data, generated from the widespread use of consumer neurotechnology applications, will provide more and more material on the basis of which conversation-stopping, neural explanatory claims can be put forward. In addition, these claims will serve to justify an expansion of brain data collection and processing as a means of providing more and more interesting claims about what those very data can be used to explain. The strange picture emerges that as consumers of BCI devices, users' brain recordings are processed into databases on which analytics are operated. From repeated data processing on brain databases, means of suggesting neural explanations for various phenomena come to the fore. These explanations then serve to promote excitement about ourselves in terms of our brains. We then expect more such excitement from further neurotechnology uses, and the cycle continues. In that our brain data furnish the databases from which everything flows, we are both the source of the advances that enable the excitement that then entices us as consumers to participate in marketing our processed brain data back to us. This kind of strange scenario can emerge whether or not genuine insights about the brain could emerge from brain data collected *en masse*.

An additional consequence could emerge if it was indeed possible to leverage troves of brain data through reprocessing to gain genuine insights on the brains of consumers. A future could open up in which neural states become the primary target for marketing, and political messaging, rather than individual persons or groups. Just as some of the hope for psychiatry as clinically applied neuroscience came from making a Karl Jaspers-inspired gestalt in data to represent a person, in a wider context than psychiatry such an impulse could signal a drive toward neuroprofiling and targeting brain data gestalts over persons.

If the bedrock of neural explanation could be hit and it really did explain human thought and behaviour, then finding ways to prompt conducive brain states could replace efforts to convince people to choose one way or another based on expressed preferences. Targeting information at specific individuals or groups based on their neural data would represent a new front in data-driven marketing or political campaigning, enabling novel forms of manipulation (Ienca et al., 2018; Kellmeyer, 2018). Corporations might market products based on brain-based profiles of consumers, bypassing choice as much as possible. Police and security services could seek to prevent criminality and terrorism by identifying ill intent through brain types. Political parties might develop policy based not on expressed voter preference, but brain-based predictions about hidden desires (See Westen et al., 2006).

An article in the New York Times presented a scene not unlike this in anticipation of the 2008 US presidential candidate debates. Using neuroscientific jargon, the candidates were described in terms of the hidden, neural reactions prospective voters had to viewing pictures of each candidate, with these reactions being cited as explanation for the eventual choices made in parties' selection of candidates. This idea of finding neural bases for political persuasion remains live, with a research group in 2022 claiming that,

> "... liberals and conservatives have noticeable and discriminative differences in functional connectivity that can be identified with high accuracy using contemporary artificial intelligence methods and that such analyses complement contemporary models relying on socio-economic and survey-based responses." (Yang et al., 2022)

While it's not at all clear whether functional connectivity patterns produce political leaning, or political preference shapes brain connectivity, for one thing, the drive for a neuroreductionist understanding of political ideology ought not to be scoffed at. As described above, in response to Tversky and Kahneman's selection task experiments, Gerd Gigerenzer was able to make apparent cognitive errors disappear through re-phrasing experimental questions in decision tasks, based in an appreciation of the evolutionary basis of decision-making. Another dimension to this kind of understanding would be trying to evoke behaviour in target audiences based in exploitation of their brain functioning. We don't have access to the specifics of our own brain functioning. But if someone else does, and has profiled us on the basis of it, they might seek to produce cognitive outcomes the way Gigerenzer sought to diminish errors. Gigerenzer made it easier for research participants to produce the expected outcomes in experiments by changing task descriptions into more amenable formats. If brain data could shine a light on neural processing types, then trying to promote 'desirable' decision-making in individuals might be simplified through leveraging knowledge about those processing types—through the use of types of evidence, spurs, prompts, etc. This might be an effective way to provoke pro- attitudes in political contexts, or become a more efficient way to sell goods than would advertising and leaving things to mere choice. Whether it can be said to accurately reflect a story of political ideology becomes less important if such insights might be instrumentalised for some other means.

All of this ought to prompt reflection. Not only would the prospects for errors make this a concern, but underlying it all is a change in the focus of how we deal with people in reducing them to brain data. We ought to think carefully about the use of data that could demonstrate ways to hack decision-making such that wider ideas of freedom of choice might be impinged upon. Whether we ought to accept such a reduction is ultimately an ethical question.

## Ethics Concerns: Ownership, Use, Markets in Data

Data has been described as 'the new oil' in some contexts, as a way to suggest its centrality for an emerging world shaped and optimised by data (The Economist, 2017). Oil has retained a particular power-conferring status for a very long time, certainly throughout the C20th. Oil is not just an energy source, and one capable of being refined into varieties of fuels but is also essential for the production of plastics and other materials. In a sense, whoever has power over oil supply has power *simpliciter*. The comparison between oil and data suggests something of the magnitude of emerging data markets and reliance upon data. Much of the infrastructure on which nations are run is digital. More and more of everyday living constitutes a 'socio-technical system.'

Socio-technical systems are systems that include technical and institutional artefacts, processes, and human actors working together to realise goals. Think of aviation, for example. The aviation sector in constituted by socio-technical systems including aeroplanes, aviation laws, land and air crews, airport security, flight corridors and controllers, among other elements. Thinking about how a plane can get from A to B involves considering a variety of these elements all coordinated according to reliable principles. With more and more complex socio-technical systems, requiring faster and more global-scale coordination, data becomes an essential part of them. Data supplies the matter and the form of global aviation coordination. Data is 'the new oil' at least in the sense that without it, much of what we are used to as normal in globalised contexts of trade, travel, and cooperation ceases to be possible. Whoever has power over data, on the analogy, has power *simpliciter*. There is one particular distinction between the two, however. Oil isn't *about* anything—data is. If I lose a barrel of oil, I can replace it by finding another barrel of oil. The loss of data is harder to replace because what data can represent is what makes it valuable.

When it comes to brain data, this idea of value being related to representation has a complicating factor. Much of the reason for discussing brain data at all comes from the discourse emerging from apprehension about novel neurotechnologies. Novel devices, and associated techniques of neural signal recording and data processing, may by some be taken to mean the mind and its contents are made vulnerable in an unprecedented manner thanks to a new dawn in rendering the brain legible (Rose, 2016). The dangers in new neurotechnologies appear to reside mainly in the provenance of novel technologies as growing from technology companies, and the lifeblood of new systems in terms of data. A disregard for respectful handling of personal data has marked many of the technology companies we are familiar with, like Facebook, Google, Amazon, and Twitter. A wholesale datafication of every dimension of technology-users' lives seems part and parcel of engaging with these and other companies. With a move into the neural, a new front is opened.

For some, there is nervousness about what brain data could reveal—maybe through processing brain data personal information could be revealed (Ienca et al., 2018). For some, there is fear that Big Tech's influence could give brain data undue influence in profiling individuals, or swaying policies, and create a new example of what Cathy O'Neil describes as a 'weapon of math destruction' based in subcranial goings on. Yet others are concerned that a data economy is a bad thing anyway, and giving over yet more data to the existing irresponsible players simply erodes privacy and cedes more power to those with too much already (Véliz, 2021).

Data is the correct focus for concerns relating to novel neurotechnologies, processing methods, and Big Tech players. This is partly because worries about specific attacks on, or threats to, the mind are overblown. Increased knowledge of neurofunctionality does not translate trivially among different individuals, nor provide reliable extra-clinical means of mental prediction or intervention (Klein, 2010; Dwyer et al., 2018). This is partly owing to a paucity in the vocabulary and ontology surrounding mental states in the first place—what we say about what brain states are is not consistent so it doesn't represent something easily translatable from one context to another (Brown, 2006; Poldrack, 2011). Interventions on the mind are an ambition of neurotechnology developers, but these ambitions are not based in generating understanding of a general model of mind and neurodynamics as much as they are on making predictions about neurofunction that *presumably* have consistent effects on mental states (Eickhoff & Langner, 2019). Specifically, these predictions are based in neural data processing. In this sense, an account of mental integrity from Andrea Lavazza seems relevant:

> "Mental Integrity is the individual's mastery of his mental states and his brain data so that, without his consent, no one can read, spread, or alter such states and data in order to condition the individual in any way." (2018, p. 4)

The reading of data regarding the brain and making predictions about mental content on the basis of that data could compete with an individual's 'mastery' of their own mental states. Not being based in rational or discursive norms, this kind of intervention would have no established place in a 'received wisdom' of mental mastery. If we are talking to another person about our mental happenings, we can decide how seriously to take their input based on a variety of factors like their level of knowledge, their disposition toward us, as well as the coherence of what they say. This is missing for a new dawn of brain data derived predictions about brains and their associations with presumed mental effects. How authoritative are brain data, for instance: Should a prediction be believed because it is based in complex data science? Could it be better at predicting mental states than a person? These could be questions in need of real consideration.

Thinking again about how brain data are derived from brain activity, we can recall how neuroelectrical signals can be recorded by electrodes placed, for example, on the scalp. These recordings include a variety of signals, superimposed on one another from different areas of the brain. In order to get useful data, algorithmic sorting of the signals is carried out. Valuable data is thus derived from noisy signals. Depending on purposes, different data might be sought. Research applications might look for specific data relating to task-relative neural activation. A medical application might look for irregular electrical activity characteristic of epilepsy. A consumer application might seek to isolate data relevant to some definition of 'meditative state' or to present a visualisation of brain activity. In such cases, the data derived from the signal might be about how brain and cognition correlate, about health and illness, or about neurofeedback-driven self-control.

In processing brain recordings in these sophisticated ways, a means is provided of predicting fundamental relations between the brain and the mind, or the person more broadly. Rather than the small, lab-based, focussed studies of the sort described above, a future of vast brain recording databases processed algorithmically could lead to far-reaching predictions about our minds. In a future of widespread consumer neurotechnology use, brain data could be collected on a huge scale beyond the lab or the clinic. What's more, as algorithms are trained on greater and greater datasets, and they develop accordingly, the possibility of reprocessing signals for novel purposes emerges too.

Signals recorded for one purpose might conceivably be sorted according to different algorithms at different times, for novel purposes. The reason we ought to consider such things now, despite there being no real prospect yet of mind reading, takes us back to algorithms. Because of the ways many algorithms learn, they increase in versatility the more data they have to operate on. Where very large datasets come into play, algorithms can detect patterns at a scale not possible for individual, or teams of, human beings. And they can do it at speeds beyond human capacity. This is why we already see AI being used to screen for cancers more reliably than some doctors, and to predict novel effective drugs (Fleming, 2018). But they can also be difficult to guide. In fact, the point of many applications is to let algorithms loose on huge amounts of unsorted data in order to see what they can discern from within the apparent data chaos.

Especially where large technology companies have access to brain data as derived from their products, troves of that data will be aggregated and maintained by those companies. Learning from the context of Big Data, it can be predicted that where complex, unstructured data sets are present, AI processing is not far behind (Gitelman, 2013). This kind of processing is automatic, often with no guiding theory. Algorithms can be simply let loose on complex, unstructured data in order to reveal patterns which may not be revealed otherwise. Given the growing sophistication of data processing, there is no way to predict what might be revealed through automated, artificially intelligent processing of large amounts of human brain data. With such processing let loose on troves of privately held consumer brain data, derived from consumer devices, the status of brain data—personal, medical, or something else—is raised.

Brain data might very easily be seen as personal data insofar as it might be used to identify and predict things about people. From general cases of data processing at scale and the re-identification of individuals from supposedly de-identified datasets it has been estimated that "…99.98% of Americans would be correctly re-identified in any dataset using 15 demographic attributes." (Rocher et al., 2019). In neurotechnological applications, recordings can be processed by algorithms aimed at predicting brain-related activity, including its relation to mental states (Poldrack, 2011). Especially when combined with contextual data, brain data may reveal highly private information. EEG recordings of electrical activity across large areas of the brain can be processed in order to diagnose diseases, like epilepsy, but increasingly also to find neural correlates of wakefulness, attention and other distinct brain states. This could provide a means of identifying individuals and ascribing to them neural or mental characteristics, beyond making inferences about their locations or lifestyle choices. For those with access to it, including tech companies like Google, Neuralink, and Facebook, and data broker firms like Equifax, Epsilon, and Experian, massed brain data will be a valuable asset and commodity.

Facebook's 2017 proposal of a brain-computer interface for "brain typing", i.e. writing text through the wilful modulation of brain activity, would not only record and process brain signals, but also associate the data derived from them with detailed social media activity. This would allow neuroprofiling, with brain data appearing to offer unvarnished and intimate insights into a person's behaviour and cognition. This could provide fertile grounds for intimate predictions about people, their behaviour, preferences, and the connections they have with others, especially in conjunction with all the other data types that are being constantly harvested by these companies. Online, this includes clicks, 'likes', retweets, expressed or inferred preferences, web history, page scroll rates, membership of or centrality in networks, gaze, location, etc. Even offline, companies like Cuebiq provide location information on those deemed to engage with billboard advertisements, based on mobile phone metadata. According to the European General Data Protection Regulation, 'personal data' is any data from which a person might be identified without unreasonable effort. On this definition, brain data look like they should qualify like many of these other types. If it did qualify as personal data, then as such it ought to be considered as property of the individual themselves rather than as a resource available to a private company without conditions.

Given also the possibility for deriving medical information from brain data, and the open possibilities for brain signal reprocessing toward novel ends, maybe brain data ought to be considered medical data. If it were to be so considered, it would attract considerable legal protection owing to its sensitivity. If a consumer neurotechnology device became popular, and routinely recorded and stored troves of consumers' brain data in cloud servers belonging to a company, this data would need different levels of consent if conceived of as medical.

For instance, re-use of the data in order to optimise future algorithms might be considered misuse of medical data. If it was collected for the purposes of, say, a neurofeedback meditation application then it might only be used for that singular purpose. Certainly, to aggregate medical data from many people in a single database would require explicit informed consent. But where signal reprocessing might occur in these contexts via algorithms let loose on unstructured datasets, consent for reprocessing would be difficult to achieve owing to the black box effect of the processing itself. It might not be known what the purpose of the processing was (it might not have a 'purpose,' being not hypothesis- but data-driven). The form of consent that might be best suited to this context is 'broad consent,' wherein one consents to any re-use of data. But given the potentially sensitive nature of possible outcomes, this could sit uncomfortably.

Related to consent is the idea that individuals ought to have the ability for no-consequence withdrawal of their data at any time. Given the complexity of the data ecosystems that brain data would appear in, this seems very complicated. For instance, if the very algorithms evolving through training on growing datasets are themselves drawing on and deriving new data from new and existing signals, how does one go about separating the data from the algorithm? If I were to have my data removed from a database, it may nevertheless along with all other data in that database already have contributed to the current version of a specific algorithm. Or, it may have featured in the production of a brain model now used to make predictions about brain/mind correlations. How would my data be meaningfully 'withdrawn' having already served a formative purpose in such model construction? In these two connected issues about data withdrawal an issue emerges about just how private brain data might be.

It seems that, although each person's brain data are their own in some sense, *any* person's brain data can impact on *everyone's* brain data. The algorithms and models derived from troves of processed brain data become at once personal and impersonally generic. They are personal as they belong in at least some sense to the individual from whom they are derived, and they can ground predictions about that very person. But they are impersonal and generic in that they serve to make processes and models that can be used to make predictions about anyone at all. The problems of how, meaningfully, to remove one's own brain data from vast troves of brain data illustrate this. Again, in terms of consent, it becomes slightly hollow to refuse consent for one's own brain data to be used in developing algorithms and models if significant numbers of others acquiesce. The algorithms and models go beyond the data into generic prediction and so the idea of private data becomes difficult to maintain.

As the processing of brain data intensifies, the profiling of neurotechnology users based on brain data is likely, given the interests and current business models of tech companies. Market forces cannot be relied upon to regulate this activity, especially not where the possibility exists for changes in ownership and the associated changes in data ownership that could result. Events beginning in April 2022 surrounding the sale of Twitter to Elon Musk serve well to highlight how agreements to terms and conditions for data collection, storage, and use can be disrupted when a company changes hands. Nevertheless, shifting the responsibility for dealing with these complex issues in human-neurotechnology-AI interactions to the individual is not a viable solution either, given the issues just described. There is increasing public understanding about online data collection and use. Nevertheless, there remains much public complacency, and ongoing misuse in business, of that data. Brain data represent a new front in this area, in need of specific ethical, legal and regulatory attention.

Computer processing opens up huge possibilities for processing the data derived from brain signals, gaining insights into the brain, and grounding speculation about the mind. Most of our brain processes are involuntary and largely automatic, serving to maintain physical and cognitive equilibria about which we are scarcely aware. As well as issues of consent in providing brain data, there is also one of neural 'data exhaust', akin to the digital trails left as we browse the web (Zuboff, 2019). Visiting websites can provide data on a user's web history, their location, their general associations with other users. Predictions about individuals can be made from web data exhaust; so too from neural data exhaust. With access to signal sorting algorithms itself largely unregulated, anyone with access to brain data could themselves carry out 'home brew' brain data processing like the contexts just described albeit on a smaller scale. Nevertheless, right now there is no real prospect of a neuroscientist, algorithm, device, or hobbyist being able to read your mind through using brain data, outside of very narrow applications, and under very specific laboratory-controlled conditions. And even then, what happens is not truly 'mind reading' but rather 'brain reading plus auxiliary hypotheses'. But as long as good predictions can be made, where 'good' can simply mean *impressive*, we need to consider the emerging possibilities that brain data insecurity could be seen to infringe upon the hitherto private realm of the mind. For example, hiring and firing in the workplace might come to include neuroprofiling in a not-too-distant future. In fact, it is already available with services presenting themselves as "…your personal HR neuro-specialist. Through a brain-model interface, users can review their unique neuroanatomical cognitive profile." This service provider goes on to say, "Our unique state-of-the-art technology can reveal *in the most objective way* all the essential elements you need your employees to possess." [emphasis added] (*Neuroprofiling for HR—A Noesis Neuromarketing product: Home*, no date). What presents itself here, with the possibility of services like this, is the potential for belief to outstrip truth.

As long as neurotechnology might come to be believed to be a silver bullet in revealing fundamental truths about people, it might start to be taken up and used as such regardless of how true the claims made about it turn out to be. This would represent a scenario in which brain data were held to outstrip people's claims about themselves, and thereby to replace testimony with neurostatistical reductionism as the basis for interpersonal assessment. In the job market, in this instance, but one could see similar movements for university admissions, loan applications and mortgages, maybe even friendship and romance. A sort of social credit system based in brainhood could emerge regardless of the validity of its basis in neuroscience, maybe just because in a complicated world many seek simplifying explanations. And how much more basic and fundamental an explanation of a person can one get than the activity of their very brain… right?

Such oddly dystopian outcomes are really only possible on the basis that brain data are not closely considered before they emerge as a widespread commodity. This would occur as a result of a growth in consumer neurotechnology. On the one hand, the emergence of brain data onto the scene ought only to happen in a way that reflects existing values widely held about personal data, the rights of the individual to mental privacy, and other established norms about how we interact, deal with, and evaluate one another. These norms have emerged over a long period and, while no doubt imperfect, they reflect interpersonal standards derived from complex social interaction. Where established norms don't have a history in interpersonal, fair discussion, we ought in general to replace them with norms that do have the legitimacy gained from that genesis. Emerging technologies have the potential to disrupt these established norms and processes of re-establishing legitimate norms, through appearing all of a sudden and with the apparent power to bring revolutionary insights. In the case of neurotechnology, the power comes via brain data.

This isn't to say, however, that technology innovation ought to be held back simply to accommodate the ways things are. Rather, the integration of novel technologies ought to, at least in the beginning, respond to established values in order to situate it in a way that allows it to be understood. Presented with a novel *next big thing* one might be tempted to see a step change in how to think of the world. Brain data might be seen to fill this role in apparently offering unvarnished personal insights in 'the most objective way.' But if it is to be seen in this way, it should be as a result of careful consideration. The responsibilities for this can't all come from society at large—it is too burdensome to expect everyone to carry out personal technology assessment exercises to establish the likely impacts of novel disruptive technology. Nor would it be satisfactory to simply have experts explain why brain data do or do not fulfil their promises. Expert opinion and established social values or norms gained through time fulfil different functions. Their influence upon one another ought to be examined in a wider discussion about what is wanted from science, technology, and society.

Considerations such as these, and the specific issues concerning brain data including how they are conceived, consent for their collection and processing, and the complex nature of their privacy in a broader algorithmic or AI context, all suggest private interests are too narrow to deal with the emergence of a market in brain data. No one entity—society (somehow conceived), technology developers and companies, policymakers for instance—has the authority, information, or capacity to speak once and for all on every important dimension of the possibilities raised through brain data emerging from a consumer context. They each have something to say, each likely have predictable biases, and each have important knowledge to contribute. It's broadly true that we don't want to live in societies closed to scientific advances. It's also the case that we don't want societies bound to the will of a hypothetical scientific ruling class, nor a free-wheeling cadre of *tech bros*. We don't want to be tied to the past, nor do we want to glibly dispense with established wisdom in light of purported innovation. This all being so, when we are presented with an emerging field of consumer neurotechnology we need to consider carefully the kinds of topics raised here.

Are the best representations of ourselves based in accessing hidden realms of data, or should we work on the basis of what people say and do? The former can be presented as unvarnished reality, but also includes involuntary brain activity, and requires interpretation via algorithm and by other people. What we say and do, meanwhile, represents things we have decided to do on reasons of our own. Interpretation is always required, but in this instance it centres on reasoned expression. If brain data don't supply a silver bullet to everything that's promised by *neuro-hype*, is it nevertheless important and due careful attention? Perhaps large-scale aggregation and an emerging market isn't appropriate for such data, at least until deeper knowledge of the entire neurotechnology sector is far more widespread publicly. Better knowledge facilitates better decision making. But the mere presence of a novel technology is not in itself sufficient to create the demand that new knowledge is sought. Without a generalised interest having emerged socially to gain such knowledge, novel technology may justifiably remain marginalised. This might be frustrating to enthusiasts and developers, but there is no social duty to adapt created by the sheer presence of a technological innovation.

# Bibliography

Bae, Y. *et al.* (2017) 'Automated network analysis to measure brain effective connectivity estimated from EEG data of patients with alcoholism', Physiological Measurement, 38(5), pp. 759–773. doi:https://doi.org/10.1088/1361-6579/aa6b4c.

Beyerstein, B.L. (1990) 'Brainscams: Neuromythologies of the new age', International Journal of Mental Health, 19(3), pp. 27–36.

Blain, L. (2019) Review: Hands-free flight with EEGSmart's mind-controlled UDrone, New Atlas, 8 March. Available at: https://newatlas.com/udrone-mind-controlled-drone-umind-review/58791/. Accessed 24 Apr 2020.

Brown, R. (2006) 'What is a brain state?', Philosophical Psychology, 19(6), pp. 729–742. doi:https://doi.org/10.1080/09515080600923271.

Buzsáki, G. and Watson, B.O. (2012) 'Brain rhythms and neural syntax: Implications for efficient coding of cognitive content and neuropsychiatric disease.', Dialogues in Clinical Neuroscience, 14(4), pp. 345–367.

Dwyer, D.B., Falkai, P. and Koutsouleris, N. (2018) 'Machine learning approaches for clinical psychology and psychiatry', Annual Review of Clinical Psychology, 14(1), pp. 91–118. doi:https://doi.org/10.1146/annurev-clinpsy-032816-045037.

Eickhoff, S.B. and Langner, R. (2019) 'Neuroimaging-based prediction of mental traits: Road to utopia or Orwell?', PLoS Biology, 17(11), p. e3000497. doi:https://doi.org/10.1371/journal.pbio.3000497.

Fleming, N. (2018) 'How artificial intelligence is changing drug discovery', Nature, 557(7707), pp. S55–S57. doi:https://doi.org/10.1038/d41586-018-05267-x.

Gitelman, L. (ed.) (2013) 'Raw Data' is an oxymoron. Cambridge, MA; London, UK: The MIT Press.

Ienca, M., Haselager, P. and Emanuel, E.J. (2018) 'Brain leaks and consumer neurotechnology', Nature Biotechnology, 36, pp. 805–810. doi:https://doi.org/10.1038/nbt.4240.

Kellmeyer, P. (2018) 'Big brain data: On the responsible use of brain data from clinical and consumer-directed neurotechnological devices', Neuroethics Preprint. doi:https://doi.org/10.1007/s12152-018-9371-x.

Klein, C. (2010) 'Images are not the evidence in neuroimaging', The British Journal for the Philosophy of Science, 61(2), pp. 265–278. doi:https://doi.org/10.1093/bjps/axp035.

Kramer, A.D.I., Guillory, J.E. and Hancock, J.T. (2014) 'Experimental evidence of massive-scale emotional contagion through social networks', Proceedings of the National Academy of Sciences, 111(24), pp. 8788–8790. doi:https://doi.org/10.1073/pnas.1320040111.

McCabe, D.P. and Castel, A.D. (2008) 'Seeing is believing: The effect of brain images on judgments of scientific reasoning', Cognition, 107(1), pp. 343–352. doi:https://doi.org/10.1016/j.cognition.2007.07.017.

Poldrack, R.A. (2011) 'Inferring mental states from neuroimaging data: From reverse inference to large-scale decoding', Neuron, 72(5), pp. 692–697. doi:https://doi.org/10.1016/j.neuron.2011.11.001.

Privacy International (2019) REPORT: Your mental health for sale. Available at: http://privacyinternational.org/report/3193/report-your-mental-health-sale. Accessed 4 Sept 2019.

Rocher, L., Hendrickx, J.M. and de Montjoye, Y.-A. (2019) 'Estimating the success of re-identifications in incomplete datasets using generative models', Nature Communications, 10(1), p. 3069. doi:https://doi.org/10.1038/s41467-019-10933-3.

Rose, N. (2016) 'Reading the human brain: How the mind became legible', Body & Society, 22(2), pp. 140–177.

The Economist. (2017). The world's most valuable resource is no longer oil, but data, 5 November. Available at: https://www.economist.com/leaders/2017/05/06/the-worlds-most-valuable-resource-is-no-longer-oil-but-data. Accessed 26 Apr 2022.

Véliz, C. (2021) Privacy is power. London: Bantam Press.

Weisberg, D.S. et al. (2007) 'The seductive allure of neuroscience explanations', Journal of Cognitive Neuroscience, 20(3), pp. 470–477. doi:https://doi.org/10.1162/jocn.2008.20040.

Westen, D. et al. (2006) 'Neural bases of motivated reasoning: An fMRI study of emotional constraints on partisan political judgment in the 2004 US presidential election', Journal of Cognitive Neuroscience, 18(11), pp. 1947–1958.

Yang, S.-E. et al. (2022) 'Functional connectivity signatures of political ideology', PNAS Nexus, p. pgac066 doi:https://doi.org/10.1093/pnasnexus/pgac066.

Zuboff, S. (2019) The age of surveillance capitalism: The fight for a human future at the new frontier of power. Paperback edn. London: Profile Books.

# Data on the Brain

**Abstract** A central contention that has been evolving throughout this book is the idea that where brain data grow in volume and use, challenges can emerge regarding how to make an account of oneself, or of persons in general. Brain data promise that such accounting can be derived from recording the activity across many instances, generalised and processed as a source presumed to disclose otherwise hidden truths. This has been connected with a wide variety of themes, including a general enthusiasm for neuroscientific discovery through computational means, the potential for research into brains to open new horizons in self-understanding, clinical revolutions regarding neurology and psychiatry, and excitement about novel brain-controlled devices and applications. An abiding theme throughout all of this has been the potential for brain data to become a kind of commodity in an obscure market, that itself can become re-commodified as the basis for algorithmic and AI advances.

**Keywords** Personal identity • Mind-reading • Social norms • Data-profiling • Public policies

S. Rainey, *Philosophical Perspectives on Brain Data*,
https://doi.org/10.1007/978-3-031-27170-0_5

It has been observed throughout this book that at least two issues ought to concern us about developments in brain data. Firstly, if brain data really are potentially disclosive of sensitive information about persons—be it about their mind, their dispositions, or their identity—then those data ought to be very carefully controlled. Not least because we can expect private companies to drive much of the likely neurotechnology innovation. If these data really can indicate insights to dimensions of our humanity like identity, mental states, or dispositions, there are considerable risks regarding its collection, storage, and legitimate use. Secondly, if brain data cannot actually deliver revolutionary insights to the degree anticipated in some quarters, it may yet be believed to deliver, with the possibility for affecting social, personal, and institutional arrangements. In this case, it seems that brain data ought to attract careful regulation too, at the very least the provision of demystifying strategies to assert the real possibilities attaching to it beyond *neurohype*.

The nub of this problem is that the agency of a person contributes much to how we typically understand their place in the world, as a conscious, active being. This is also the best basis we have for treating them as valuable and demanding of respect. From any individual perspective, each of us can ordinarily offer accounts of what '*I*' means in terms of ourselves. Whatever other concepts and ideas we have and entertain, '*I*' can be used meaningfully. Self-ignorance, self-deception, self-alienation, total dissociation are all possibilities, where *I* is uttered in a way disconnected from the person. But these are not the norm. Yet concepts like these lie behind the idea that revelations can come from drawing back a veil of superficial behaviour to reveal the irrational or other hidden dimensions of human beings. When experiments are carried out that seem to show hidden sources and motivations for behaviour, we might have pause for thought or to consider how in control we really are. But these considerations are interesting and worthwhile only to the extent that we already understand the meaningful use of accounting for ourselves in '*I*' terms. To prompt a revision of this general group of ideas or variations on its themes would require a great deal of clear evidence for an improved account. It isn't clear that this improved account came from the sorts of investigations carried out by Tversky and Kahneman, and nor is it clear that it might come from reductionism to brain data.

Understanding the brain, and its connections with how we think and behave, includes areas of research with growing technological sophistication. Research here may be capable of providing tantalising insights into hitherto rather arcane-seeming brain processes. This research might give clues to associations among brain dynamics and mental phenomena including mental images, imagined speech, and dreaming. Growing understanding of the brain allows for ambitious predictions to be made about mental contents in very controlled circumstances. Nevertheless, there is no genuine near-term prospect of mind-reading of the sort often seen in science-fiction. Mental contents remain out of reach to neuro-technologies, and brain data don't bring them closer into reach. Detailed predictions about the brain, derived from processing brain data, don't provide the kinds of difference-making accounts we expect from an account that would characterise an action as one thing or another—the significance of an action is opaque to brain data, just as it is to an over-zealous physical reductionism of the mental to the neural. The best way to discover truths about someone's mental states remains, I maintain, to talk with them.

But there remain mind-reading risks from the advance of neurotech-nologies. This is especially centred upon how data processing works in technologies developed outside of clinical and scientific research. Neurotechnologies will begin amassing ever growing databases of brain recordings. Technology companies will continue developing algorithms capable of discerning more than meets the human eye among data. In combination, these will amount to a great many opportunities for pre-dicting things about brains and inferring things about minds. Such a sophisticated, techno-scientific approach to illuminating important dimensions of the person will likely be powerful enough to rival some standard accounts of brains and minds for some people. We need to think now about whether this is something we want to undergo, or something over which we should take control.

## Reasonable Expectations

The classic motto *festina lente*—make haste slowly—could serve generally for scientific advances that bring technological innovation. This seems certainly the case where brain data are at stake. In order not get caught up in *neurohype*, and to maintain reasonable expectations for the complex area of neurotechnology innovation, it ought not to be assumed that fast progress is good progress. Moreover, it ought not to be assumed that advance is necessarily progress in a more evaluative sense. Advances might include a range of moves whereby some technology or process becomes streamlined or made more accessible to specific groups of people. Within the confines of a specific practice or discipline this might mark a progressive move. For instance, wider access to neural signal sorting algorithms might make for technical progress in neuroimaging as it opens the field for greater numbers of smaller players to produce useful results. But 'progress' here doesn't translate automatically into progress *per se*. Should this kind of progress lead to more powerful consumer BCIs, those BCIs may yet provide little of genuine benefit or interest to users. Different groups will have different values at stake, and so different evaluations of what counts as advance, progress, beneficial, risky, and so on should be expected.

In ongoing monitoring of changes in neurotechnological fields, there ought to be a means of including values in a forum for discussion open to the possibility that, should advances be possible within neurotechnology development, established values may become strained. In this context, themes triggered by *neurohype*-filled statements could be useful to open questions about what's at stake with brains, minds, data, the self, wellness and illness, and so on. Following detailed evaluative conversations, established values could perhaps be justifiably revised. The justification of value revision, should it arise, ought to be carefully taken on as a task, rather than assumed to be inevitable in the face of technological innovation. Likewise, should technological innovation demand value revision, concerning views of the self or means of identifying health and illness, it ought to be recognised that this demand is not in and of itself decisive. Since these technologies are about people, and sometimes revolve around commodifying dimensions of those peoples' experience through the development of a market in brain data, people ought to be very much central to considerations of how valuable the technological innovation itself may be.

Brain data offer possibilities for greater neuroscientific understanding of brain functioning. This includes its maladies, and the cognitive or behavioural problems such maladies can bring. This could revolutionise mental healthcare in particular, leading to highly targeted diagnostic and treatment strategies. But the ways in which brain data will be acquired, processed, and used raise questions. How much we can trust AI systems to read our brains, and how much we understand the sorts of 'reading' those systems undertake have ramifications for ways we ought to think of mental health and illness. The neuron doctrine, placing the neuron at the centre of understanding the brain as inaugurated in the work of Santiago Ramón y Cajal especially, has been a powerful guiding notion in neuroscience. In the sense that, through data, it comes to matter less which neuron fired when, this doctrine could come to mean less.

Where data fuels understanding of the brain through a discovery science approach, knowledge of brain processes and correlations with mental states could emerge. But without a clear, material connection to the brain, and a perhaps hard to follow algorithmic genesis, how certain ought we to be of gaining an understanding of these processes and correlations? We ought not to be overawed by the sheer achievement of scientific knowledge but remember that we have choices in how to evaluate and use scientific discovery. We have stakes in science, for sure, especially when there is the promise of effective and efficient healthcare. But we have stakes in other dimensions of living too, including ideas of scientific rigour, personal identity, civic freedoms and duties, and ethics.

The collection and use of brain data outside research and clinical settings especially ought to prompt scientifically informed, socio-political dialogue, rather than straightforward consumer-industrial adoption. We know from other fields that where there has been industrial-scale data collection, problems have followed. These have included questions about ownership, consent, and legitimate use. We have the scientific and multidisciplinary expertise to convene far-reaching discussions on what new technologies offer, and what they risk. We have the means for inventive ways to enact research-society engagement. Prudence suggests we ought to sail a course between the extremes of eagerness and skepticism, in order that we remain in control of our health, our selves, our freedoms, and our data. This can be achieved through careful consideration, open dialogue, and anticipatory public policymaking. How we mediate and moderate scientific activity and technological innovation in social discourses will reflect not only on our ability to do science, but also how to wield it well.

## Social Norms

Whilst the possibility of damaging, sensitive, or unwelcome inferences being drawn about people based on their brain data has been raised in this book, there have also been doubts about the veracity of such inferences. Nevertheless, claims made about people needn't be true to cause concern, and those based in brain data are likely to be taken very seriously whether true or not, such is the allure of the neural. But there is another, wider issue raised with the prospect of widespread brain data collection, processing, sale, and use. The socio-political dimensions of a market in brain data could bring issues we would be wise to forestall. Given the loose, unregulated nature of widespread brain signal capture made possible by consumer BCIs in particular, vast databases could come into being that will be assets for processing by any means deemed useful or interesting by tech companies and related parties.

Just as with data brokering based in other data sources—mobile phone metadata, internet search histories, purchases made—these will be processed, packaged, and sold as profiles of types of person. The aim of such profiling may be to market goods and services through targeted advertisements, to stimulate political activity as with tailored political messaging, or perhaps in the name of security through producing *neurotypes* of risky people—if we can predict psychiatric disease, after all, why not criminality, giving us an edge on who to monitor for likely future unrest.

But this entire field of data-profiling is untested. Though it can be seen to get results by some metrics—political messaging seems to become more effective through microtargeting, for instance—it isn't clear *why* or *how* it gets results. Once the neuro- is added to the sets of data available, the prestige of these profiles will grow, again, owing to the draw the brain has. But this has the effect that a novel socio-political and economic tool for monitoring and influencing citizens may come into effect without oversight or validation. This makes a laboratory of society itself, instrumentalising the lives of communities in pursuing private, data-led, endeavours to change decision-making, values, and norms across the public. Given the interests of those involved, the scale and complexity of the effort, and the opacity of each step, it isn't clear how this public-as-data-science lab could be assented to, nor opposed. This brings up questions about how 'society as a lab' appears, especially how science and technology innovation might resist criticism in such a context.

Scientific activity is not always very open to questions from outside science. Scientific progress might be thought of by many of those participating in it as being bound primarily by norms of rigour, and values of pursuing truths with scant thought of cost. In a context of future challenges on a global scale that might be met only through careful scientific advance, the public sphere in general somewhat inevitably becomes a kind of 'collective experimentation' laboratory. But considering areas such as climate change, air pollution, strains on water supplies, these require large scale responses that have international, multi-jurisdictional, implications for lots of different people, with lots of different views of the world. This being so, these many publics require—at least in principle—some kind of channel of response to science. This response might not be in the form of a scientific proposition or counterclaim to a hypothesis. But where changes to modes of life are tabled, other types of input become more vital for a rounder view of 'progress'.

The idea of the public as a sort of laboratory may appear to be quite problematic. Normal labs have funders, aims, stated goals. What is the research agenda of the public as a lab? Is it open that individuals and groups might opt out of that lab's experiments? Normal labs running social science experiments, for example, need to be very clear about the nature of their studies, offering information to participants, with clear inclusion and exclusion criteria. What would this look like for a public as laboratory. Labs like those found in universities have to gain ethics approval for their exploratory work. What would such approval look like when the lab was the public itself, and the research agenda was unstated? Without structures like those present as constraints on normal lab research the public as a lab is ethically and scientifically questionable. Ethically, as approval for its activities looks dubious, and scientifically as its goals are unstated, methods unclear, and its research base vaguely specified.

It may be that there is no effective arena for critical discussion of the public as lab besides the courts. This cannot be expected to suffice, given the complexity of such a context. Leaving things to the courts, apart from the *ad hoc* and retrospective nature of such a solution, also means that the society as lab continues unopposed until a crisis emerges. This produces an asymmetry among societies (not to mention the internal complexities of 'societies') and those whose interests are served by the collective experimentation. Where technology development is at stake, this pits big tech against *we the people*. In the case of brain data, this dynamic seems clear. Neurotechnology producers, data processing and brokering entities, and wider related markets will gain from widespread brain data collection and use. But what remains to be seen is a solid argument why societies ought to go along for the ride.

Where a discursive basis for the introduction of a new technology into the public sphere is missing, where solid arguments aren't forthcoming, there is the potential for unjustified social change. In the case of brain data as with many others, this represents change predicated on narrow sectoral interests. Constructing a discursive basis for potential social change is a necessary condition for a just public sphere. Such justifications will stem from communicative action rather than experimental ambition. The legitimacy of a scientifically-informed social change does not come solely from its basis in good science. It must also somehow be considered more generally as 'good' by those who must be motivated to change their behaviour in response to it. The formulation, processes, and communication of policy are thus vital, raising questions in terms of legitimacy, alongside effectiveness (Habermas, 1980). This minimally involves accounting for, and communicating appropriately, the different justifications for social changes and for their acceptance. But this can only be effective if justifications for social changes are legitimate and trustworthy. This means that some types of 'non-scientific', discursive evidence must be expected to feature in the overall evaluation of scientific progress likely to yield social change.

When norms or values clash in these sorts of circumstances, often what's at stake is one conception of 'the good life' or another. Where accepting, or changing, norms is prompted by one set of choices rather than another, it is people who must reorient themselves in a new reality. Such reorientation can be positive, but where it is handled badly or, worse, ignored, it represents domination of one set of views by another. How the good life is envisaged; how one ought to recognise others different to oneself; when ought change to be encouraged; these are clearly important areas for careful reflection. Scientific progress, sometimes out of step with other conceptions of progress, raises all kinds of concerns. It seems clear that science that hopes to change the world, or the lives of others, must engage with those affected in a responsive dialogical manner.

## Public Policies

Brain data exist among an array of fields in which scientific progress and technological innovation might clash with what people think of as 'progress' more generally. The focus here ought to be on how science and society talk to each other, how they co-exist and relate to one another. Such matters ought to be thought of in terms of the ethical, social, political, and economic questions they raise—and the discursive opportunities they present—rather than the legal verdicts they could elicit following some crisis event. With brain data, we are currently in a position where too much progress has not yet been made, and there still exists an opportunity to step in and respond to the potentialities opened by them.

In collecting large amounts of data from many users, brain recording techniques generate large troves of brain data that will permit population-level predictions about more people than just those participating in research. This raises neural data ownership, misuse and re-use concerns, at a population level as well as that of the individual. The possibility of reprocessing existing brain signal recordings with evolving algorithmic and AI approaches is available owing to the nature of the recording, on the one hand, whereby much more signal than a given purpose required is recorded by an EEG. Predictions about individuals can be made from brain data exhaust in ways that may not be anticipated owing to the potential for AI's inclusion in this reprocessing. These are so far under-considered potentialities arising from brain signal recording and (re)processing of brain data.

From existing and ongoing research and criticism of AI in a variety of applications, we already know that data doesn't serve all equally. Human biases in data selection, and in choosing topics on which to gather data, mean datasets always reflect past practice. Data are historical in this sense. When predictions are made about people based in brain data, those predictions will be historically based, in data sets not representative of all, and with hidden interests represented by obscure data collection rationales. Yet as persons, as workers, as citizens, we ought not to expect our futures to be shackled to past actions. No matter how sophisticated the predictions made may be, we ought to be alert to the possibility that they presume the past is a good model for the future. Ironically in this context, such a backward-looking mode of thinking comes in the garb of progress, innovation, and *disruptive technology*. The past might give good insights for the present and the future, but the lessons to learned from it come from how well we interpret what we learn: what we ask of the past and why. This relates once more to agency and how it is exercised.

As long as any technology, or practice with a technology, challenges the exercise of agency there ought to be scrutiny applied. Brain data in research, clinical, and consumer contexts can present such an agency-stifling technology in presenting a deterministic picture of persons, offering a map to the future based in analyses of the past. The ways in which predictions about persons based in their brain data might emerge in opposition to their own expressed aims, or their self-conception, ought not to be *presumed* as advancing science. The split brain setup suggested it was possible in certain circumstances to know more about a person's situation than they themselves might. But subsequent discussion of conceptual limits in prediction and the value of agency should provide pause in taking this too far. Added to this, there is the possibility that the kinds of profiling carried out based in brain data would classify people by neurotypes rather than preferred or other more familiar categorisations. This could serve to undermine social norms and legal protections regarding those preferred or familiar means of self-classification.

Moreover, when we think of ourselves as citizens and how state power gains legitimacy (in functioning democracies at any rate, in the ideal) there are longstanding assumptions about our capacity to make free and fair decisions concerning who will lead and govern us. This socio-political view of persons grates with one wherein we are predictable based in analyses of brain data. If we can't be said to act deliberately on reasons, to have voted in elections following due consideration, but instead to simply act out ensembles of behaviour latent in dynamic patterns of neural activation, then we lose agency personally and socio-politically. The New York Times discussion from 2017, relating neural reactions to candidates, represents a fundamental challenge to the power relation between the governed and the governing. If policies were to be aimed at groupings of neurotypes, these may not map onto any given set on citizens constituting an electorate. The neurotypes, wrought from algorithmic data processing, need not map onto sets of natural persons, since they represent machine classified general neurotypes and predictions from these to preferences. Where do the citizens who would vote for their preferred candidates come in?

The suggestion here is that when personal agency and political legitimacy are on the scales, brain data and their applications would need to be very weighty to tip the balance. As of now, the ambitions some hold for a brain data revolution lack the seriousness of what they would challenge. Neurotechnology ought to be regulated so as to determine the parameters within which it can safely develop according to solid scientific norms, in line with well-considered socio-political values, and can be explained and marketed honestly to consumers. The data from neurotechnology should be regulated carefully in terms of the devices in whose functioning it will operate. How practices, such as psychiatry, ought to use data of this kind ought to be a matter for professional bodies regulating those practices. In the widest sense, a strong regime of data protection ought to be readied for brain data's emergence.

Current data protection, such as the EU's General Data Protection Regulation (GDPR) is likely insufficient to deal with the novelty and pace of developments in brain data (European Commission, 2018;

Rainey et al., 2020). The GDPR is aimed at making personal data more secure in the sense that individuals are given more control over the data derived from them by companies, institutions, or other entities. But the regulation categorises data according to how and for what purpose it is recorded. If a neural device is intended as a neural controller for a piece of hardware or software, relevant neural signals can be extracted from neural recordings in order to trigger, control and optimise that device or application. However, other information could be derived from the same recordings of those signals by means of subsequent reprocessing and interpreted in ways that make its status difficult to determine.

Where consumer devices recordings might be processed so as to produce potentially medically relevant predictions, these may be *sensitive* data that ought to attract special protection. This is perhaps especially the case given the rate at which recording density is increasing, the greater understanding of how inter-neuron communication affects information processing and the likely increased future role for machine learning in neural data analysis (Stevenson & Kording, 2011). But these kinds of eventualities do not sit easily in the wording of the regulation. The emerging context of consumer devices is not anticipated by the GDPR, nor is the nature of the data it can produce. The stakes, moreover, are not simply the potential identification of people from one time to another—though research into 'brain fingerprinting' is ongoing (see for instance Gui et al., 2014; Bashar et al., 2016; Sorrentino et al., 2022). In the context of brain data and a potential market in their predictive uses, the stakes include personal agency, and socio-political legitimacy.

Without greater public understanding of neurotechnological developments, guided by good scientific communication, regulated by forward thinking policy, there exists a potential minefield when it comes to brain data. Without a constructive, well-informed and regulated context in which to emerge, brain data may appear with insufficient safeguards and produce a gold rush among powerful companies and institutions. While it is always possible that social norms might evolve spontaneously in light of scientific and technological advance, we ought to be wary of that potential evolution being driven by the sheer presence of technology and the hype surrounding it. Brain data have the potential to become the basis for neurotechnological advance in research, clinical, and consumer contexts. But moves toward these ends ought to progress in proportion with public understanding of the stakes, not sectoral calculations of market advantage or professional prestige.

# Bibliography

Bashar, Md.K., Chiaki, I. and Yoshida, H. (2016) 'Human identification from brain EEG signals using advanced machine learning method EEG-based biometrics', in 2016 IEEE EMBS conference on biomedical engineering and sciences (IECBES), Kuala Lumpur, Malaysia: IEEE, pp. 475–479. doi:https://doi.org/10.1109/IECBES.2016.7843496.

European Commission (2018) The GDPR: New opportunities, new obligations. Brussels. Available at: https://ec.europa.eu/commission/sites/beta-political/files/data-protection-factsheet-sme-obligations_en.pdf. Accessed 30 Mar 2019.

Gui, Q., Jin, Z. and Xu, W. (2014) 'Exploring EEG-based biometrics for user identification and authentication', in Proceedings of IEEE signal processing in medicine and biology SPMB, pp. 1–6.

Habermas, J. (1980) *Legitimation crisis*. Translated by T. MacCarthy. London: Heinemann.

Rainey, S. *et al.* (2020) 'Is the European data protection regulation sufficient to deal with emerging data concerns relating to neurotechnology?', Journal of Law and the Biosciences, 7(1), p. lsaa051. doi:https://doi.org/10.1093/jlb/lsaa051.

Sorrentino, P. *et al.* (2022) 'Brain fingerprint is based on the aperiodic, scale-free, neuronal activity'. bioRxiv, p. 2022.07.23.501228 doi: https://doi.org/10.1101/2022.07.23.501228.

Stevenson, I.H. and Kording, K.P. (2011) 'How advances in neural recording affect data analysis', Nature Neuroscience, 14(2), pp. 139–142. doi:https://doi.org/10.1038/nn.2731.

# Bibliography

Amunts, K. *et al.* (2016) 'The human brain project: Creating a European Research Infrastructure to Decode the Human Brain', Neuron, 92(3), pp. 574–581. doi:https://doi.org/10.1016/j.neuron.2016.10.046.

Anderson, C. (2008) 'The end of theory: The data deluge makes the scientific method obsolete', Wired, 16(07).

Andrew Butterfield, Gerard Ekembe Ngondi, and Anne Kerr (eds) (2016) A dictionary of computer science. 7th edn. Oxford: Oxford University Press.

Bae, Y. *et al.* (2017) 'Automated network analysis to measure brain effective connectivity estimated from EEG data of patients with alcoholism', Physiological Measurement, 38(5), pp. 759–773. doi:https://doi.org/10.1088/1361-6579/aa6b4c.

Bashar, Md.K., Chiaki, I. and Yoshida, H. (2016) 'Human identification from brain EEG signals using advanced machine learning method EEG-based biometrics', in 2016 IEEE EMBS conference on biomedical engineering and sciences (IECBES), Kuala Lumpur, Malaysia: IEEE, pp. 475–479. doi:https://doi.org/10.1109/IECBES.2016.7843496.

Bashashati, A. *et al.* (2007) 'A survey of signal processing algorithms in brain–computer interfaces based on electrical brain signals', Journal of Neural Engineering, 4(2), p. R32. doi:https://doi.org/10.1088/1741-2560/4/2/R03.

© The Author(s), under exclusive license to Springer Nature Switzerland AG 2023

S. Rainey, *Philosophical Perspectives on Brain Data*,

https://doi.org/10.1007/978-3-031-27170-0

Baughman, R.W. *et al.* (2006) 'The National Institutes of Health Blueprint for Neuroscience Research', Journal of Neuroscience, 26(41), pp. 10329–10331. doi:https://doi.org/10.1523/JNEUROSCI.3979-06.2006.

Bethlehem, R.A.I. *et al.* (2022) 'Brain charts for the human lifespan', Nature, 604(7906), pp. 525–533. doi:https://doi.org/10.1038/s41586-022-04554-y.

Beyerstein, B.L. (1990) 'Brainscams: Neuromythologies of the new age', International Journal of Mental Health, 19(3), pp. 27–36.

Blain, L. (2019) Review: Hands-free flight with EEGSmart's mind-controlled UDrone, New Atlas, 8 March. Available at: https://newatlas.com/udrone-mind-controlled-drone-umind-review/58791/. Accessed 24 Apr 2020.

boyd, Danah and Crawford, K. (2012) 'Critical questions for big data', Information, Communication & Society, 15(5), pp. 662–679. doi:https://doi.org/10.1080/1369118X.2012.678878.

Brown, R. (2006) 'What is a brain state?', Philosophical Psychology, 19(6), pp. 729–742. doi:https://doi.org/10.1080/09515080600923271.

Buzsáki, G. and Watson, B.O. (2012) 'Brain rhythms and neural syntax: Implications for efficient coding of cognitive content and neuropsychiatric disease.', Dialogues in Clinical Neuroscience, 14(4), pp. 345–367.

Bzdok, D. and Meyer-Lindenberg, A. (2018) 'Machine learning for precision psychiatry: Opportunities and challenges', Biological Psychiatry: Cognitive Neuroscience and Neuroimaging, 3(3), pp. 223–230.

Caruso, G.D. (2021) 'On the compatibility of rational deliberation and determinism: Why deterministic manipulation is not a counterexample', The Philosophical Quarterly, 71(3), pp. 524–543. doi:https://doi.org/10.1093/pq/pqaa061.

Changeux, J.-P. (2017) 'Climbing brain levels of organisation from genes to consciousness', Trends in Cognitive Sciences, 21(3), pp. 168–181. doi:https://doi.org/10.1016/j.tics.2017.01.004.

Changeux, J.P., Courrège, P. and Danchin, A. (1973) 'A theory of the epigenesis of neuronal networks by selective stabilization of synapses', Proceedings of the National Academy of Sciences of the United States of America, 70(10), pp. 2974–2978.

Cheour, M. *et al.* (2001) 'Mismatch negativity and late discriminative negativity in investigating speech perception and learning in children and infants', Audiology and Neurotology, 6(1), pp. 2–11. doi:https://doi.org/10.1159/000046804.

Choudhury, S. and Slaby, J. (2016) Critical neuroscience: A handbook of the social and cultural contexts of neuroscience. Wiley.

Churchland, P.S. (1989) Neurophilosophy toward a unified science of the mind brain. MIT Press.

Coates McCall, I. *et al.* (2019) 'Owning ethical innovation: Claims about commercial wearable brain technologies', Neuron, 102(4), pp. 728–731. doi:https://doi.org/10.1016/j.neuron.2019.03.026.

Conrad, P. and Barker, K.K. (2010) 'The social construction of illness: Key insights and policy implications', Journal of Health and Social Behavior, 51(1_suppl), pp. S67–S79. doi:https://doi.org/10.1177/0022146510383495.

Dalvie, S. *et al.* (2016) 'Toward a global roadmap for precision medicine in psychiatry: Challenges and opportunities' OMICS: A Journal of Integrative Biology, 20(10), pp. 557–564. doi:https://doi.org/10.1089/omi.2016.0110.

Davidson, D. (2001) Essays on actions and events. 2nd edn Oxford; New York: Clarendon Press; Oxford University Press.

Deco, G. *et al.* (2019) 'Awakening: Predicting external stimulation to force transitions between different brain states', Proceedings of the National Academy of Sciences, 116(36), pp. 18088–18097. doi:https://doi.org/10.1073/pnas.1905534116.

Dinov, I.D. (2016) 'Volume and value of big healthcare data', Journal of Medical Statistics and Informatics, 4(1), p. 3. doi:https://doi.org/10.724 3/2053-7662-4-3.

Dwyer, D.B., Falkai, P. and Koutsouleris, N. (2018) 'Machine learning approaches for clinical psychology and psychiatry', Annual Review of Clinical Psychology, 14(1), pp. 91–118. doi:https://doi.org/10.1146/annurev-clinpsy-032816-045037.

Eickhoff, S.B. and Langner, R. (2019) 'Neuroimaging-based prediction of mental traits: Road to utopia or Orwell?', PLoS Biology, 17(11), p. e3000497. doi:https://doi.org/10.1371/journal.pbio.3000497.

Eronen, M.I. (2020) 'Interventionism for the intentional stance: True believers and their brains', Topoi, 39(1), pp. 45–55. doi:https://doi.org/10.1007/s11245-017-9513-5.

European Commission (2018) The GDPR: New opportunities, new obligations. Brussels. Available at: https://ec.europa.eu/commission/sites/beta-political/files/data-protection-factsheet-sme-obligations_en.pdf. Accessed 30 Mar 2019.

Farahany, N.A. (2011) 'A neurological foundation for freedom', Stanford Technology Law Review.

Fleming, N. (2018) 'How artificial intelligence is changing drug discovery', Nature, 557(7707), pp. S55–S57. doi:https://doi.org/10.1038/d41586-018-05267-x.

Gamez, D. (2014) 'Are information or data patterns correlated with consciousness?', Topoi, pp. 1–15.

Gazzaniga, M.S. (1998) 'The split brain revisited', Scientific American, 279(1), pp. 50–55.

Gazzaniga, M.S. (2005) 'Forty-five years of split-brain research and still going strong', Nature Reviews Neuroscience, 6(8), pp. 653–659. doi:https://doi.org/10.1038/nrn1723.

Gigerenzer, G. (1991) 'How to make cognitive illusions disappear: Beyond "Heuristics and Biases"', European Review of Social Psychology, 2(1), pp. 83–115. doi:https://doi.org/10.1080/14792779143000033.

Gitelman, L. (ed.) (2013) 'Raw Data' is an oxymoron. Cambridge, MA; London, UK: The MIT Press.

Goering, S. and Klein, E. (2020) 'Fostering neuroethics integration with neuroscience in the BRAIN initiative: Comments on the NIH neuroethics roadmap', AJOB Neuroscience, 11(3), pp. 184–188. doi:https://doi.org/10.1080/21507740.2020.1778120.

Gui, Q., Jin, Z. and Xu, W. (2014) 'Exploring EEG-based biometrics for user identification and authentication', in Proceedings of IEEE signal processing in medicine and biology SPMB, pp. 1–6.

Habermas, J. (1980) Legitimation crisis. Translated by T. MacCarthy. London: Heinemann.

Hardcastle, V.G. and Stewart, C.M. (2002) 'What do brain data really show?', Philosophy of Science, 69(S3), pp. S72–S82. doi:https://doi.org/10.1086/341769.

Hogarth, S., Javitt, G. and Melzer, D. (2008) 'The current landscape for direct-to-consumer genetic testing: Legal, ethical, and policy issues', Annual Review of Genomics and Human Genetics, 9(1), pp. 161–182. doi:https://doi.org/10.1146/annurev.genom.9.081307.164319.

Horikawa, T. et al. (2013) 'Neural decoding of visual imagery during sleep', Science, 340(6132), pp. 639–642. doi:https://doi.org/10.1126/science.1234330.

Huth, A.G. et al. (2016) 'Natural speech reveals the semantic maps that tile human cerebral cortex', Nature, 532(7600), pp. 453–458. doi:https://doi.org/10.1038/nature17637.

Ienca, M., Haselager, P. and Emanuel, E.J. (2018) 'Brain leaks and consumer neurotechnology', Nature Biotechnology, 36, pp. 805–810. doi:https://doi.org/10.1038/nbt.4240.

Insel, T.R. (2014) 'The NIMH research domain criteria (RDoC) project: Precision medicine for psychiatry', American Journal of Psychiatry, 171(4), pp. 395–397. doi:https://doi.org/10.1176/appi.ajp.2014.14020138.

Insel, T.R. and Quirion, R. (2005) 'Psychiatry as a clinical neuroscience discipline', JAMA, 294(17), pp. 2221–2224. doi:https://doi.org/10.1001/jama.294.17.2221.

Ismael, J.T. (2016) How physics makes us free. Oxford University Press doi:https://doi.org/10.1093/acprof:oso/9780190269449.001.0001.

Kellmeyer, P. (2018) 'Big brain data: On the responsible use of brain data from clinical and consumer-directed neurotechnological devices', Neuroethics Preprint. doi:https://doi.org/10.1007/s12152-018-9371-x.

Kirmayer, L.J. and Crafa, D. (2014) 'What kind of science for psychiatry?', Frontiers in Human Neuroscience, 8. doi:https://doi.org/10.3389/fnhum.2014.00435.

Kirmayer, L.J. and Gold, I. (2011) 'Re-socializing psychiatry: Critical neuroscience and the limits of reductionism', in S. Choudhury and J. Slaby (eds) Critical neuroscience. Oxford, UK: Wiley-Blackwell, pp. 305–330. doi:https://doi.org/10.1002/9781444343359.ch15.

Kitchin, R. (2014) 'Big data, new epistemologies and paradigm shifts', Big Data & Society, 1(1), p. 2053951714528481. doi:https://doi.org/10.1177/2053951714528481.

Klein, C. (2010) 'Images are not the evidence in neuroimaging', The British Journal for the Philosophy of Science, 61(2), pp. 265–278. doi:https://doi.org/10.1093/bjps/axp035.

Kramer, A.D.I., Guillory, J.E. and Hancock, J.T. (2014) 'Experimental evidence of massive-scale emotional contagion through social networks', Proceedings of the National Academy of Sciences, 111(24), pp. 8788–8790. doi:https://doi.org/10.1073/pnas.1320040111.

Kringelbach, M.L. et al. (2020) 'Dynamic coupling of whole-brain neuronal and neurotransmitter systems', Proceedings of the National Academy of Sciences, 117(17), pp. 9566–9576. doi:https://doi.org/10.1073/pnas.1921475117.

Kringelbach, M.L. and Deco, G. (2020) 'Brain states and transitions: Insights from computational neuroscience', Cell Reports, 32(10). doi:https://doi.org/10.1016/j.celrep.2020.108128.

Kulkarni, S. et al. (2020) 'Artificial intelligence in medicine: Where are we now?', Academic Radiology, 27(1), pp. 62–70. doi:https://doi.org/10.1016/j.acra.2019.10.001.

Lavazza, A. (2018) 'Freedom of thought and mental integrity: The moral requirements for any neural prosthesis', Frontiers in Neuroscience, 12. Available at: https://www.frontiersin.org/article/10.3389/fnins.2018.00082. Accessed 9 Feb 2022.

Lélé, S. and Norgaard, R.B. (2005) 'Practicing interdisciplinarity', Bioscience, 55(11), pp. 967–975 doi:https://doi.org/10.1641/0006-3568(2005)05 5[0967:PI]2.0.CO;2.

Levy, N. (2007) Neuroethics. Cambridge; New York: Cambridge University Press.

Lotte, F. *et al.* (2018) 'A review of classification algorithms for EEG-based brain–computer interfaces: A 10 year update', Journal of Neural Engineering, 15(3), p. 031005. doi:https://doi.org/10.1088/1741-2552/aab2f2.

Marino, S. *et al.* (2018) 'Controlled feature selection and compressive big data analytics: Applications to biomedical and health studies', PLoS One, 13(8), p. e0202674. doi:https://doi.org/10.1371/journal.pone.0202674.

Markram, H. *et al.* (2011) 'Introducing the human brain project', Procedia Computer Science, 7, pp. 39–42. doi:https://doi.org/10.1016/j.procs.2011.12.015.

Mayer-Schönberger, V. and Cukier, K. (2013) Big data: A revolution that will transform how we live, work, and think. Houghton Mifflin Harcourt.

McCabe, D.P. and Castel, A.D. (2008) 'Seeing is believing: The effect of brain images on judgments of scientific reasoning', Cognition, 107(1), pp. 343–352. doi:https://doi.org/10.1016/j.cognition.2007.07.017.

McCrone, P.R. (ed.) (2008) Paying the price: The cost of mental health care in England to 2026. London: King's Fund.

Mitchell, R.J., Bishop, J. and Low, W. (1993) Using a genetic algorithm to find the rules of a neural network. doi:https://doi.org/10.1007/978-3-7091-7533-0_96.

*Muse—meditation made easy muse.* (n.d.). Available at: https://choosemuse.com/. Accessed 18 Dec 2019.

*Neuroprofiling for HR—a noesis neuromarketing product: Home.* (n.d.). Available at: http://neuroprofiling.eu/index.html. Accessed 20 Apr 2022.

Nishimoto, S. *et al.* (2011) 'Reconstructing visual experiences from brain activity evoked by natural movies', Current Biology, 21(19), pp. 1641–1646.

O'Neill, C. (2016). *Weapons of math destruction, how Big Data increases inequality and threatens democracy.* Penguin.

Peirce, C.S.S. (1906) 'Prolegomena to an apology for pragmaticism', The Monist, 16(4), pp. 492–546.

Pernu, T.K. (2018) 'Mental causation via neuroprosthetics? A critical analysis', Synthese, 195(12), pp. 5159–5174. doi:https://doi.org/10.1007/s11229-018-1713-z.

Place, U.T. (1956) 'Is consciousness a brain process?', British Journal of Psychology, 47(1), pp. 44–50.

Poldrack, R.A. (2006) 'Can cognitive processes be inferred from neuroimaging data?', Trends in Cognitive Sciences, 10(2), pp. 59–63.

Poldrack, R.A. (2011) 'Inferring mental states from neuroimaging data: From reverse inference to large-scale decoding', Neuron, 72(5), pp. 692–697. doi:https://doi.org/10.1016/j.neuron.2011.11.001.

Privacy International (2019) REPORT: Your mental health for sale. Available at: http://privacyinternational.org/report/3193/report-your-mental-health-sale. Accessed 4 Sept 2019.

Pycroft, L. *et al.* (2016) 'Brainjacking: Implant security issues in invasive neuro-modulation', World Neurosurgery, 92, pp. 454–462. doi:https://doi.org/10.1016/j.wneu.2016.05.010.

Rainey, S. *et al.* (2020) 'Is the European data protection regulation sufficient to deal with emerging data concerns relating to neurotechnology?', Journal of Law and the Biosciences, 7(1), p. lsaa051. doi:https://doi.org/10.1093/jlb/lsaa051.

Rainey, S. (2022) 'Datafied brains and digital twins: Lessons from industry, caution for psychiatry', Philosophy, Psychiatry, & Psychology, 29(1), 29–42. doi:https://doi.org/10.1353/ppp.2022.0005.

Rainey, S. and Erden, Y.J. (2020) 'Correcting the brain? The convergence of neuroscience, neurotechnology, psychiatry, and artificial intelligence', Science and Engineering Ethics, 26(5), pp. 2439–2454. doi:https://doi.org/10.1007/s11948-020-00240-2.

Regier, D.A., Kuhl, E.A. and Kupfer, D.J. (2013) 'The DSM-5: Classification and criteria changes', World Psychiatry, 12(2), pp. 92–98. doi:https://doi.org/10.1002/wps.20050.

Roache, R. and Savulescu, J. (2018) 'Psychological disadvantage and a welfarist approach to psychiatry', Philosophy, Psychiatry, & Psychology, 25(4), pp. 245–259. doi:https://doi.org/10.1353/ppp.2018.0035.

Rocher, L., Hendrickx, J.M. and de Montjoye, Y.-A. (2019) 'Estimating the success of re-identifications in incomplete datasets using generative models', Nature Communications, 10(1), p. 3069. doi:https://doi.org/10.1038/s41467-019-10933-3.

Rose, N. (2010) '"Screen and intervene": Governing risky brains', History of the Human Sciences, 23(1), pp. 79–105. doi:https://doi.org/10.1177/0952 695109352415.

Rose, N. (2016) 'Reading the human brain: How the mind became legible', Body & Society, 22(2), pp. 140–177.

Rose, N. and Abi-Rached, J. (2014) 'Governing through the brain: Neuropolitics, neuroscience and subjectivity' The Cambridge Journal of Anthropology, 32(1), pp. 3–23. doi:https://doi.org/10.3167/ca.2014.320102.

Schölkopf, B. (2015) 'Learning to see and act', Nature, 518(7540), pp. 486–487. doi:https://doi.org/10.1038/518486a.

Schultze-Lutter, F., Schmidt, S.J. and Theodoridou, A. (2018) 'Psychopathology—a precision tool in need of re-sharpening', Frontiers in Psychiatry, 9, p. 446. doi:https://doi.org/10.3389/fpsyt.2018.00446.

Smart, J.J. (1959) 'Sensations and brain processes', The Philosophical Review, pp. 141–156.

Sorrentino, P. et al. (2022) 'Brain fingerprint is based on the aperiodic, scale-free, neuronal activity'. bioRxiv, p. 2022.07.23.501228 doi: https://doi.org/10.1101/2022.07.23.501228.

Stevenson, I.H. and Kording, K.P. (2011) 'How advances in neural recording affect data analysis', Nature Neuroscience, 14(2), pp. 139–142. doi:https://doi.org/10.1038/nn.2731.

The Economist. (2017). The world's most valuable resource is no longer oil, but data, 5 November. Available at: https://www.economist.com/leaders/2017/05/06/the-worlds-most-valuable-resource-is-no-longer-oil-but-data. Accessed 26 Apr 2022.

Toga, A.W. et al. (2015) 'Big biomedical data as the key resource for discovery science', Journal of the American Medical Informatics Association, 22(6), pp. 1126–1131. doi:https://doi.org/10.1093/jamia/ocv077.

Tversky, A. and Kahneman, D. (1983) 'Extensional versus intuitive reasoning: The conjunction fallacy in probability judgment', Psychological Review, 90(4), p. 23.

Véliz, C. (2021) Privacy is power. London: Bantam Press.

Vul, E. et al. (2009) 'Puzzlingly high correlations in fMRI studies of emotion, personality, and social cognition', Perspectives on Psychological Science, 4(3), pp. 274–290. doi:https://doi.org/10.1111/j.1745-6924.2009.01125.x.

Weisberg, D.S. et al. (2007) 'The seductive allure of neuroscience explanations', Journal of Cognitive Neuroscience, 20(3), pp. 470–477. doi:https://doi.org/10.1162/jocn.2008.20040.

Westen, D. *et al.* (2006) 'Neural bases of motivated reasoning: An fMRI study of emotional constraints on partisan political judgment in the 2004 US presidential election', Journal of Cognitive Neuroscience, 18(11), pp. 1947–1958.

Wodak, R., Chilton, P.A. and Wodak-Chilton (eds) (2005) A new agenda in (critical) discourse analysis: Theory, methodology, and interdisciplinary. Amsterdam: Benjamins Discourse approaches to politics, society, and culture, 13.

Wright, J. (2018a) 'Chapter 11—seeing patterns in neuroimaging data', in C. Ambrosio and W. MacLehose (eds) Progress in brain research. Elsevier Imagining the Brain: Episodes in the History of Brain Research, pp. 299–323. doi:https://doi.org/10.1016/bs.pbr.2018.10.025.

Wright, J. (2018b) 'The analysis of data and the evidential scope of neuroimaging results', The British Journal for the Philosophy of Science, 69(4), pp. 1179–1203. doi:https://doi.org/10.1093/bjps/axx012.

Yang, S.-E. *et al.* (2022) 'Functional connectivity signatures of political ideology', PNAS Nexus, p. pgac066 doi:https://doi.org/10.1093/pnasnexus/pgac066.

Zuboff, S. (2019) The age of surveillance capitalism: The fight for a human future at the new frontier of power. Paperback edn. London: Profile Books.

# Index